Probabilistic Reliability Analysis of Power Systems

Bart W. Tuinema · José L. Rueda Torres ·
Alexandru I. Stefanov ·
Francisco M. Gonzalez-Longatt ·
Mart A. M. M. van der Meijden

Probabilistic Reliability Analysis of Power Systems

A Student's Introduction

Bart W. Tuinema (iD)
Section Intelligent Electrical Power Grids
Department of Electrical Sustainable Energy
Delft University of Technology
Delft, The Netherlands

José L. Rueda Torres (iD)
Section Intelligent Electrical Power Grids
Department of Electrical Sustainable Energy
Delft University of Technology
Delft, The Netherlands

Alexandru I. Stefanov
Department of Electrical Sustainable Energy
Delft University of Technology
Delft, The Netherlands

Francisco M. Gonzalez-Longatt
Department of Electrical Engineering
University College of Southeast Norway
Kongsberg, Norway

Mart A. M. M. van der Meijden (iD)
Section Intelligent Electrical Power Grids
Department of Electrical Sustainable Energy
Delft University of Technology
Delft, The Netherlands

ISBN 978-3-030-43500-4   ISBN 978-3-030-43498-4   (eBook)
https://doi.org/10.1007/978-3-030-43498-4

This Springer imprint is published by the registered company Springer Nature Switzerland AG
The registered company address is: Gewerbestrasse 11, 6330 Cham, Switzerland

Preface

This book gives an introduction to probabilistic reliability analysis of power systems. The main objective is to introduce (electrical) engineering students to this interesting research field. Many probabilistic methods have been described in other textbooks and publications before. However, it is hard to find a textbook that describes a variety of probabilistic methods while providing students with understandable examples and exercises to practice their knowledge. This book has been developed to fill that gap. By explaining probabilistic methods, describing real-life examples and providing practical exercises, students are able to thoroughly understand the topic and develop hands-on experience. The many real-life examples discussed in this book, which are based on actual case studies in the Netherlands, not only help students understand the theory, but also make this book an interesting reference for professionals working in this field. It was tried to keep a balance by discussing the basics of a range of methods while providing references to more detailed works for further reading.

To support the study process, each chapter in this book starts with an overview of learning objectives, which can be checked after reading the specific chapter. The most important concepts (normally followed by their definition) are *italicized*. In Chaps. 3–5, practical examples and case studies are described in text boxes. At the end of each chapter, questions are included to test the knowledge and skills acquired by studying the chapter. The solutions to these questions can be found at the end of this book. Basic knowledge about probability, statistics, and load flow analysis is assumed when explaining the reliability models. Appendices A and B can be consulted to refresh this basic knowledge.

This book consists of four parts. Part I gives an introduction to reliability analysis of power systems. Chapter 1 discusses the importance of (probabilistic) reliability analysis of power system and introduces the main concepts related. Chapter 2 explains how the components of the power system can possibly fail and how this could lead to a blackout of the power system. In Part II of this book, the most common models used in probabilistic reliability analysis of power systems are described in detail. Chapter 3 concentrates on reliability modeling of individual components, whereas Chaps. 4 and 5 describe the reliability models for small and large systems, respectively. Part III of this book presents three topics related to probabilistic reliability analysis of power systems. The first, probabilistic load flow analysis (Chap. 6), focuses on the modeling of uncertainties in the power system, like load behavior and variable (renewable) generation. The second topic, reliability of EHV underground cables (Chap. 7), shows how the methods discussed in the theoretical chapters can be applied to analyze the reliability of underground cables in large transmission networks. The third topic, cyber-physical power system modeling (Chap. 8), discusses how the reliability and resilience of the power system to cyber attacks become important in the future power systems. These three topics show how the theory of this book can be applied in practice and how the theory is related to other research fields. Finally, Part IV of this book is the conclusion (Chap. 9).

This book could not have been written without the help of some colleagues. I would like to thank the following persons for their help for improving this book and/or the previous versions of its material: Robert Kuik, Rob Ross, André Bossche, Simon Tindemans, and Frank Mulder. In general, I would like to thank the colleagues of TenneT TSO B.V. for sharing the information needed to perform the studies described in this book. This enabled us to include the many practical examples that illustrate the theory. I also would like to thank some M.Sc. students that I supervised during my Ph.D. research: Fengli, Niki, and Reinout. Their studies found a place in the case studies described in this book. On behalf of all the authors, I wish you success for studying the theory of this book and I hope that this book succeeds in providing the basic knowledge needed to work with enthusiasm in this research field.

Delft, The Netherlands   dr.ir. Bart W. Tuinema
November 2019

Contents

About the Authors

Bart W. Tuinema was born in The Hague, the Netherlands, on February 5, 1984. He received his B.Sc. and M.Sc. degrees from Delft University of Technology, Delft, the Netherlands, in 2007 and 2010, respectively. He completed his Ph.D. research on the reliability of power systems with respect to new developments like EHV underground cables and networks for large-scale offshore wind energy at Delft University of Technology in 2017. After completing his Ph.D., he was involved as Postdoctoral Researcher in the projects MIGRATE (Massive InteGRATion of power Electronic devices) and TSO2020 (Electric "Transmission and Storage Options" along TEN-E and TEN-T corridors for 2020), also at Delft University of Technology. He has been Teacher of the course 'Reliability of Sustainable Power Systems' for several years and has been involved in other teaching activities related to probabilistic reliability analysis of power systems.

José L. Rueda Torres received his Ph.D. degree in electrical engineering from the Universidad Nacional de San Juan, San Juan, Argentina, in 2009. From 2003 to 2005, he worked in Ecuador in the fields of industrial control systems and electrical distribution network operation and planning. From 2010 to 2014, he was Postdoctoral Research Associate in the Institute of Electrical Power Systems, University of Duisburg-Essen. He is currently Associate Professor of stability, control and optimization in the Department of Electrical Sustainable Energy, Delft University of Technology, Delft, the Netherlands. His research interests include power system stability and control, multi-energy sector coupling, power system planning, and probabilistic and artificial intelligence methods. He is Chairman of the Working Group on Modern Heuristic Optimization (WGMHO) under the IEEE PES Analytic Methods in Power Systems (AMPS) Committee. He is also Secretary of IEEE/CIGRE C4-C2.58 Joint Working Group Evaluation of Voltage Stability Assessment Methodologies in Transmission Systems.

Alexandru I. Stefanov is Assistant Professor at the Intelligent Electrical Power Grids Group at Delft University of Technology, Delft, the Netherlands. His research interests are resilience of cyber-physical energy systems and cyber security for power grids. He holds the registered professional title of Chartered Engineer (CEng MIEI) from the Institution of Engineers of Ireland. Previously, he worked as Professional Engineer in the Operations Department at ESB Networks, Distribution System Operator in Ireland. He has postdoctoral research experience and graduated his Ph.D. in power systems at University College Dublin, Ireland. He graduated his M.Sc. and B.Sc. in power systems engineering at University Politehnica of Bucharest, Romania.

Francisco M. Gonzalez-Longatt is currently Full Professor in electrical power engineering at Universitetet i Sørøst-Norge, Norway (2019). He is Former Lecturer in electrical power systems at Loughborough University, UK (2013–2019), and Senior Lecturer in electrical engineering at Coventry University, UK (2011–2013). He was as Postdoctoral Researcher in the School of Electrical and Electronic Engineering, The University of Manchester, UK (2009–2011). Formerly, he was Associate Professor and Chair of Engineering Department of Universidad Nacional Experimental Politécnica de la Fuerza Armada Bolivariana (UNEFA), Venezuela (1995–2009). He is Vice President of the Venezuelan Wind Energy Association. His main area of interest is the integration of intermittent renewable energy resources into future power system and smart grids.

Mart A. M. M. van der Meijden received the M.Sc. degree in electrical engineering from the Technical University of Eindhoven, Eindhoven, the Netherlands, in 1981. From 1982 to 1988, he was with ASEA/ABB in the field of process automation. Since the last 20 years, he has been with different Dutch utilities. He has been involved in the development initiatives as grid technology innovation, implementation of sustainable energy, new organizational strategies, and implementation of asset management. Since 2003, he has been with TenneT TSO B.V., as Innovation Manager, where he is responsible for the development of the TenneT Vision 2030 and the initial phase of the multi-energy North Sea Wind Power Hub. His research interests include system strategic planning and control, probabilistic methods, and the dynamic interaction between offshore and onshore multi-energy systems. Since 2011, he has also been Professor and Lecturer in power systems of the future in the Department of Electrical Sustainable Energy, Faculty of Electrical Engineering Mathematics and Computer Science, Delft University of Technology, Delft, the Netherlands. He has joined and chaired several national and international expert groups.

Part I
Introduction

Chapter 1
Introduction

This chapter introduces the topic of power system reliability analysis. First, it is explained why reliability analysis of power systems is of importance and what recent developments lead to an increased need to analyze this reliability with probabilistic approaches (Sect. 1.1). Then, the basic definitions of reliability and risk are discussed in Sect. 1.2. The general approach of probabilistic reliability analysis is presented in Sect. 1.3, thereby discussing the differences between deterministic and probabilistic analysis in more detail. Some available software packages and test networks will be shortly discussed as well. As the Transmission System Operator (TSO) is the main responsible party for securing the electricity supply, the various activities performed by a TSO are described in detail in Sect. 1.4. It is explained how the different activities of a TSO are performed in order to maintain a high reliability of the power system. This chapter thereby provides the basic background knowledge needed for the further chapters of this book.

Learning Objectives

After reading this chapter, the student should be able to:

- explain the importance of (probabilistic) power system reliability analysis;
- mention definitions of reliability and risk, and discuss their differences;
- describe the differences between deterministic and probabilistic analysis;
- discuss the main activities of a Transmission System Operator, the three main TSO processes, and the corresponding time horizons and time scales.

© Springer Nature Switzerland AG 2020
B. W. Tuinema et al., *Probabilistic Reliability Analysis of Power Systems*,
https://doi.org/10.1007/978-3-030-43498-4_1

1.1 Reliability of Power Systems

In the modern society, it is hard to imagine how to live without electricity. Electricity has proven to be one of the best carriers of energy for various smaller and larger applications and is therefore widely used. When the electricity supply is interrupted, the impact can be catastrophic: The production processes within factories are interrupted, the transport system is affected, and even the security within neighborhoods can be in danger. It is therefore of utmost importance that power systems be as reliable as possible.

Several recent developments put increasing stress on power systems [1]:

- The development of and increase in electricity demand;
- Structural network changes like more interconnections, replacement of conventional energy sources by intermittent renewable energy sources (RES) and increasing use of power electronics (e.g. high-voltage direct current, HVDC);
- New operating policies like the liberalization of the energy market, more (international) trading, coupled markets, and higher demand-side participation;
- Aging of the power system components.

At the same time, it is desirable to utilize power systems optimally, which means that power systems must also be as economical as possible. The combination of these developments leads to a situation in which power systems are operated closer to their maximum capabilities. New techniques and technologies can make this possible without increasing the risk of blackouts.

One example of a new technique is probabilistic reliability analysis. In the past, power systems were mainly designed using deterministic reliability analysis and deterministic reliability criteria like $N - 1$ redundancy, which states that the failure of a single component must not lead to a failure of the power system. While deterministic criteria have proven to be clear and effective, using them can result in over-dimensioned power systems. For example, studies have shown that deterministic $N - 1$ redundancy is not economical in offshore networks. With probabilistic reliability analysis, it is possible to quantify the real risk by analyzing the likelihood and impact of possible events, such that the optimal size and structure of the power system can be determined. Current developments like the large-scale implementation of renewables (like offshore wind and solar photovoltaics) increase the need for probabilistic approaches even more. The intermittent nature of these sources and the resulting variable power flow scenarios can only be modeled properly by probabilistic approaches. Probabilistic approaches can also provide more insight into cases where deterministic approaches do not show any discriminatory results. For example, the reliability of networks with traditional overhead lines and networks with underground cables can only be compared using probabilistic analysis.

1.2 Definitions of Reliability and Risk

To study the reliability of power systems, first the concept *'reliability'* must be defined well. In literature, *system reliability* is generally defined as [2, 3]:

Definition 1.1 System reliability is the ability of a system to fulfill its function.

To understand this definition, it must first be clear what a system is. Generally, a *system* is a group of components, separated from the system environment by a system boundary, working together to fulfill a system function [2]. A *component*, then, is a part of the system that is not split up into smaller parts in further analysis [4]. Furthermore, to use this definition of system reliability, it must be clear what the *function* of the system is. The most important function of a power system is to supply the load. If the load is interrupted, then the power system is not able to fulfill its function and is therefore considered unreliable. This definition of power system reliability is used in many traditional power system reliability studies, where it is usually assumed that there is one single entity that can take any necessary action to secure the electricity supply.

In the *liberalization of the energy market*, which took place in the European Union but also in several other countries over the world, production and transport of electricity were divided over different entities. This led to various actors on the electricity market: Apart from *producers* and *consumers*, there are now also *system operators, service providers, electricity companies without own production, electricity traders* and *'prosumers'*. The *system operator* is the entity responsible for the electrical grid. Among the responsibilities are the connection of *customers* (which can be *consumers* and *producers*), the development and maintenance of the network, and the facilitation of electricity trading [5]. The power grid can be divided over multiple system operators. The *Transmission System Operator (TSO)* is then responsible for the higher-voltage part of the network, while *Distribution System Operators (DSOs)* are responsible for the lower-voltage networks. For example, in the Netherlands there is one TSO (TenneT) for the *extra high-voltage (EHV: 380/220* kV*)* and *high-voltage (HV: 150/110* kV*)* networks, while there are several DSOs (like Stedin) for the *medium-voltage (MV: 10–50* kV*)* and *low-voltage (LV: 400* V*)* networks.

The various actors on the electricity market have different views on power system reliability. Generally, consumers desire a reliable electricity supply, while producers want to be able to produce electricity and to sell this to their customers. The main tasks of the TSO are to connect all customers and to facilitate electricity trading. This also means that the function of the power system changes slightly. Whenever customers (load or generation) are interrupted, or whenever the market is hindered (e.g. by transport restrictions), the system is not able to fulfill its function and can be regarded as unreliable. Of course, the impact of load interruptions is generally much more severe than the interruption of generation or hindrance of the market, but it can be concluded that because of the change of its function, power system reliability should be seen in a broader sense nowadays.

Reliability is closely related to *risk*, defined as the product of probability and impact [6]:

Definition 1.2 Risk = Probability × Impact

Since probability is dimensionless, risk has the same dimension as the impact: if the impact is expressed in lost money, the risk is expressed in (expected) lost money. As the impact of possible events can be diverse, the Dutch TSO TenneT applies the *risk matrix* as a tool to estimate the risk of future network developments. Six *business values* are defined: *secure supply*, *engage stakeholders*, *safety*, *financial*, *environment*, and *compliance* [7]. For new developments, the risk of each business value is estimated by using the risk matrix as shown in Fig. 1.1. This provides an overview of the risks of a new development. The risk matrix assumes a small modification of the definition of risk: The impact can also be combined with the frequency of an event instead of the probability of an event. As a consequence, the risk inherits the dimension of the frequency and the impact: If the frequency is given in /year and the impact is given as lost money in euro, then the risk is the (expected) lost money in euro/year.

The previously mentioned definition of system reliability, Def. 1.1, is mainly reflected by a *secure supply*. Security of supply is a reliability indicator that reflects the reliability for the consumer. In practice, various reliability indicators exist that describe this security of supply. Security of supply does not reflect the network reliability for the producers, but this can be seen as a financial risk for the TSO and is therefore included in the business value *financial*. Some business values are directly related to the first definition of system reliability. For example, the disconnection of consumers causes a financial risk, while large blackouts will also cause risks for the business values *engage stakeholders* and *compliance*. *Safety* and *environment* are less directly related to the definition of system reliability, but this is open for discussion.

In the example illustrated in Fig. 1.1, the business values financial, secure supply, and safety have been placed into the risk matrix. The frequency of these risks can

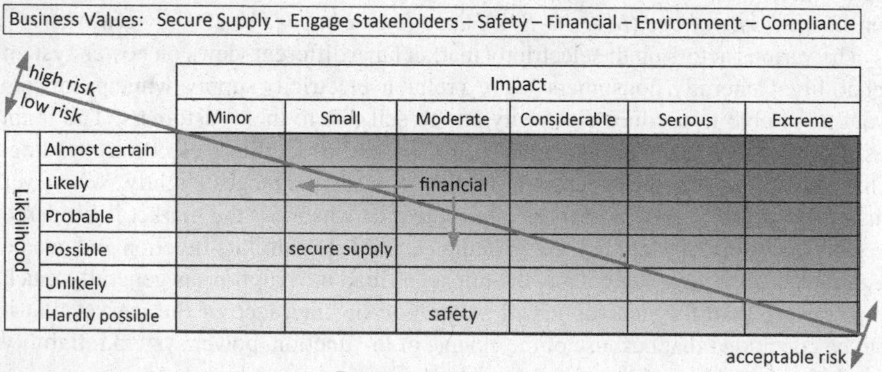

| Business Values: Secure Supply – Engage Stakeholders – Safety – Financial – Environment – Compliance |

		Impact					
		Minor	Small	Moderate	Considerable	Serious	Extreme
	Almost certain						
	Likely			financial			
	Probable						
	Possible		secure supply				
	Unlikely						
	Hardly possible			safety			

acceptable risk

Fig. 1.1 Risk matrix used by TenneT TSO

be calculated by the probabilistic approaches explained in this chapter or it can be determined by expert opinion. The impact is normally determined by case studies or experience. For each business value, an acceptable level of risk is defined. The acceptable level of risk often is a topic of discussion. In some cases, especially for financial risk, it is possible to calculate the acceptable level of risk. For other cases, the acceptable level of risk is set by experience or it is decided that the risk of the new situation must not be larger than the risk in the old situation. The example illustrated in the figure shows that the financial risk is too high in this case. As shown, this risk can be mitigated by limiting the impact or reducing its likelihood.

1.3 Reliability Analysis of Power Systems

1.3.1 Reliability Analysis Approach

In reliability analysis of power systems, it is tried to determine whether the power system is able to supply the load and fulfill its other functions. As this depends on many aspects, system reliability can be divided into *system adequacy* and *system security* [3]. *System adequacy* refers to the availability of enough facilities in the system to fulfill its function, while *system security* refers to the ability of the system to respond to faults and events occurring in the system. System adequacy therefore considers a static condition of the power system, whereas system security considers the dynamic behavior of the system. Although the methods described in this book strictly spoken belong to system adequacy, system reliability is used as the overall concept throughout this book. Furthermore, it must be realized that *security of supply* is in fact an indicator of system adequacy.

For power system reliability analysis, various approaches exist. These can be divided into *analytical* and *simulation* approaches [3]. In analytical approaches, mathematical models are developed and the reliability is calculated from the numerical results of these models. In simulation approaches, the power system is simulated with a computer and the reliability is estimated based on the system behavior during this simulation. Analytical approaches are generally preferred as simulation often requires a long simulation time. Also, analytical approaches can give more insight into the relation between the input and the results. On the other hand, simulation approaches offer the possibility to include more realistic system behavior and are preferred when analytical approaches tend to become too complicated. Apart from Monte Carlo simulation, all methods described in this book are analytical approaches.

Reliability analysis of power systems, like any analysis, follows a sequential approach of input, analysis, and output. Figure 1.2 illustrates the general approach for reliability analysis of power systems (with the focus on probabilistic reliability analysis). When preparing the input of the analysis, several topics are of importance. Data analysis (of system parameters, load patterns, wind and conventional generation) and the collection of failure statistics (like failure frequencies and repair times)

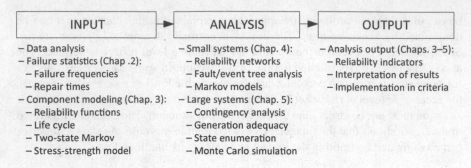

Fig. 1.2 Reliability analysis approach

are essential, as the accuracy of the results of reliability analysis depends on the accuracy of the input. Failure statistics and the process of component failure are described in detail in Chap. 2 of this book.

Next, the reliability behavior of the components within the system needs to be modeled. For this purpose, several methods exist, as mentioned in the figure. These methods are described in detail in Chap. 3 of this book. For the actual reliability analysis of power systems, a variety of approaches exists. These can be roughly divided into approaches for smaller and approaches for larger systems. Figure 1.2 mentions some of these approaches, which will be discussed in detail in Chaps. 4 and 5 of this book, respectively.

Regarding the output of the reliability analysis, an important topic is the representation of the results by clear and actionable reliability indicators. Also, the interpretation of the results into practical decisions and the implementation in new criteria that secure the reliability of power systems are of importance. These topics concerning the output will be discussed in the chapters of this book as well when describing the specific analysis methods.

1.3.2 Deterministic Versus Probabilistic Analysis

For reliability analysis of power systems, both *deterministic* and *probabilistic* approaches can be used. In *deterministic approaches*, the effects of certain events, often a set of worst-case scenarios, are studied. This is illustrated in Fig. 1.3. Each analysis of an event leads to an outcome, giving an overview of: 'If x happens, the effect is y'. It is then studied whether the effects are still acceptable in all cases and what the most severe effect is. Often, *deterministic criteria* are defined for deterministic approaches. For example, a specific component must not fail because of a typical lightning surge (as defined in standards). Another example is the $N - 1$ *redundancy* criterion, which states that a single component failure must not lead to a failure of the system. More in general, $N - \alpha$ *redundancy* means the ability of the system to operate with α components out of service.

Fig. 1.3 Deterministic versus probabilistic reliability analysis

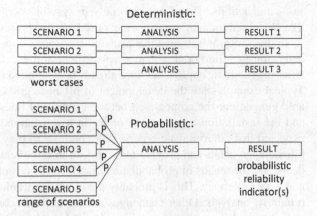

In contrast to deterministic approaches, in *probabilistic approaches* the probabilities of the events are included in the study as well. Hereby, it is possible to consider a large range of possible scenarios and events and to combine the results into one (or several) probabilistic reliability indicator(s), based on the probabilities of the input scenarios and events. This is illustrated in Fig. 1.3 as well. In general, this makes the analysis more realistic. For example, with probabilistic approaches it is possible to include the variable nature of renewable generation in the analysis, whereas deterministic approaches usually only consider several worst-case scenarios (like 'high wind—low load' or 'no wind—peak load'). Moreover, the random nature of component failures is included in the analysis and weighed according to the probability. For example, in a probabilistic approach all possible generator failures can be included, whereas deterministic approaches only consider several worst cases (e.g. 'loss of largest unit').

Although deterministic criteria are easy to understand and have proven to effectively secure power system reliability, their meaning has been topic of discussion for long, like in [8]. The main issue is that using deterministic criteria might lead to over-designed power systems, while economics and risks are neglected. For example, if a remote customer is connected, is it then economical to connect him with $N - 1$ redundancy or is he willing to accept some risk for a certain price? Is it necessary to connect offshore wind farms with $N - 1$ redundancy? Is it acceptable to allow some risk by performing maintenance under $N - 0$ in the transmission network? In all these cases, deterministic approaches do not provide sufficient insight.

Probabilistic reliability analysis is useless, however, if the results cannot be translated into actionable reliability indicators. With the many calculation methods, a variety of reliability indicators can be calculated. It is often left to the system operator to decide what values of these reliability indicators are acceptable and desirable. In reality, this decision is complicated: Why would a security of supply of 0.9999 be acceptable and a security of supply of 0.9998 not? Also, the criteria to maintain a certain reliability level in system operation are not clear. In this sense, deterministic criteria have proven to be unambiguous, effective, and easy to translate into decisions and actions. Therefore, especially in system operation, deterministic criteria like $N - 1$ redundancy are still widely used, as also prescribed in grid codes like [5].

In this sense, it is argued that probabilistic approaches should be an extension to deterministic criteria rather than a replacement of these. In the design of power systems, probabilistic approaches have already been applied successfully in many cases. Typical examples are the development of offshore grids, the integration of renewable generation, the comparison between overhead lines and underground cables, and the installation a spare transformer in a substation. In system operation, it is expected that deterministic criteria will continue to play an important role. Probabilistic approaches can provide more insight into the real risks, but in order to take decisions, the results of probabilistic reliability analysis must be presented in a clear and actionable way. This is probably the most important challenge of probabilistic reliability analysis. Other challenges are the collection of accurate input data and reducing the long computation time of probabilistic analysis of large systems.

1.3.3 Software for Power System Reliability Analysis

Over the past few decades, various software packages for (probabilistic) power system reliability analysis have been developed. Examples are *TRELSS*, *TRANSREL*, *CREAM* and *MECORE* [9]. In addition, power system analysis software like *PSS/E* and *PowerFactory* are extended with toolboxes for reliability analysis. Some of the programs are based on analytical approaches and others on simulation. This section shortly describes some of this software.

The reliability toolbox of PSS/E uses state enumeration (as explained in Sect. 5.1) to analyze the reliability of power systems [10]. In fact, PSS/E performs a deterministic multi-level contingency analysis (as explained in Sect. 5.1) up to level three. Then, as probabilistic postprocessing, the probabilities of these contingencies (i.e. component outages) are included to calculate reliability indices. The results can be presented both numerically and graphically. Like PSS/E, PowerFactory uses state enumeration for the reliability analysis [11]. The main difference with PSS/E is that PowerFactory allows one to study the system in more detail. PowerFactory can model outages within substations. Moreover, PowerFactory can model the failure behavior of system components with different models, like Weibull–Markov. Also, maintenance schedules and varying load levels can be included in the analysis.

Some programs for power system reliability analysis are specially developed for specific parts of the power system. For example, *PROMOD* analyzes the generation system, while *TPLAN* analyzes the transmission system. Other programs (like *TRELSS*, *TRANSREL*, and *CREAM*) are able to analyze the combined generation and transmission system. *TRANSREL* and *TPLAN* are based on analytical approaches, while *PROMOD* and *MARS* use Monte Carlo simulation. A detailed comparison of the variety of software is given in [9]. The choice for certain software always depends on the characteristics of the specific reliability study.

As all these programs are specially developed for power system reliability analysis, these are able to analyze the reliability of large power systems in detail. However, there are some challenges using such software. The main challenge is that detailed information (like failure behavior of all system components, load/generation scenarios, network data, control schemes, and corrective actions) is needed before reliability analysis can be performed. This information is often difficult to specify. Another challenge is that these programs cannot always be modified easily. For example, in some programs it is not possible to include load/generation scenarios or to exclude control schemes for study purposes.

An alternative is to design a reliability analysis in mathematical software like *MATLAB*. This gives the freedom to modify the reliability analysis and to present the results in the desired way. The disadvantage is that MATLAB is general mathematical software and has not specially been designed for analysis of power systems. Therefore, load flow analysis must be included into the program. Two options are to use *MATPOWER* [12], or to program a load flow specifically for the studied system.

1.3.4   Test Networks

For power system studies, several test systems have been developed. Most of these test systems were developed to study particular aspects of power systems, like transmission network planning or stability behavior. Two test systems were especially developed for reliability studies: the *RTS24* and the *RBTS*. These two are shortly described in this section.

The IEEE RTS24 (Reliability Test System 24-bus) is shown in Fig. 1.4. This test system has a realistic topology and contains various components that can also be found in real power systems, like generators/loads, overhead lines (single circuit and double circuit), underground cables, and transformers. The data for this test system is described in [14]. This also includes the component impedances/ratings, generator capacities, failure rates and repair times, and a load scenario for one year.

The IEEE RBTS (Roy Billinton Test System) was developed at the University of Saskatchewan, Canada. The RBTS is illustrated in Fig. 1.5. For educational purposes, there was a need for a simple test network that enabled researchers to gain understanding of and develop new methods for probabilistic reliability analysis of power systems. The data for this test system can be found in [15], and some basic reliability results are described in [16].

In practice, both the RTS24 and the RBTS are often used to develop, test, and demonstrate new analysis techniques. The RBTS is most useful for educational purposes. For example, a failure of line 9 directly leads to the loss of bus 6, and the generation system is not much larger than the peak load, such that load curtailment (i.e. the disconnection of load) easily occurs after a generator failure. Furthermore, the failure rates of the components are larger than they are in reality, such that the reliability indices do not become too small. The RTS24 is more realistic in this sense

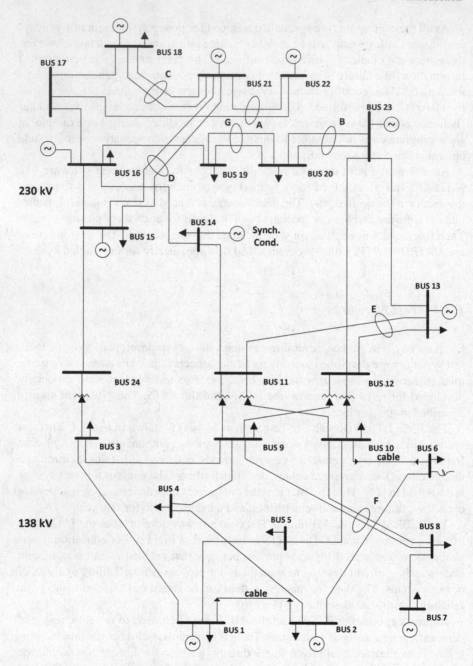

Fig. 1.4 RTS24 bus test system (adapted from [13]). ©2011 IEEE, with permission

Fig. 1.5 RBTS 6 bus test network (adapted from [13]). ©2011 IEEE, with permission

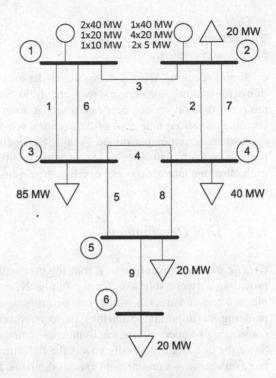

and is therefore often used to develop and demonstrate new methods for reliability analysis. A detailed survey of the use of reliability test systems can be found in [13].

The choice for a certain test system always depends on the kind of study and its purpose. For some studies, it might be best to use the RTS24 to demonstrate a new calculation approach. For other studies, it might be best to use an existing network (e.g. the Dutch EHV transmission network) to obtain more realistic results.

1.4 Transmission System Operator (TSO) Activities

1.4.1 Three Main Processes

Transmission System Operators (TSOs) perform various activities to secure a high reliability level of the power system.[1] In most TSOs, these activities are divided over the three *Transmission System Operator (TSO) processes* [17]:

---

[1]This section is based on the work published in [1]. ©2016 IET.

1. *Grid development;*
2. *Asset management;*
3. *System operation.*

Simply speaking, these processes consider the creation, maintenance, and operation of the transmission network, respectively. In each of the three processes, activities are performed on various *time horizons* (i.e. *long term*, *mid term* and *short term*), focusing on various *time scales* (like *decades*, *years*, *months*). This section describes the various activities performed by a TSO, together with the corresponding time horizons and actual time scales. As activities in different time horizons can influence each other, the interactions and overlaps are discussed as well.

1.4.2   Grid Development

Grid development mainly aims at transmission network expansion planning, thereby providing network solutions for the future. New connections and substations are planned and constructed, which can be onshore as well as offshore. For efficient planning, it is important to find the type, location, and timing of the network upgrades considering socioeconomic, environmental, reliability, legal, and political aspects. Since the fact that it generally covers the far future, grid development deals with a large number of uncertainties in various domains, like:

• When, where, and how is electricity consumed, produced, and stored?
• What is the socioeconomic impact of mega-investments on grid infrastructure?
• What are the failure rates and repair times of critical equipment and assets?
• What are the consequences of the transition toward a renewable energy supply?

It is often tried to cover these uncertainties by developing, analyzing, and comparing several *development scenarios*. These scenarios contain the size and location of generators and load in the power system, where the transition toward a renewable electricity supply is usually reflected by scenarios ranging from 'business-as-usual' to 'green revolution' [18].

In the TSO practice, grid development is performed on time horizons ranging from long term to short term. Long-term grid development considers a time scale of decades and includes the development of network expansion plans based on load/generation scenarios for the future. Examples are the development of an offshore grid and onshore network extensions. In mid-term grid development, investments in the infrastructure (like new connections and substations in the existing network topology) are made on a time scale of years to a decade. In short-term grid development, only small modifications are made in the time scale of months to years. For example, new protection systems and phase shifters can be installed in this time horizon.

1.4.3 Asset Management

Asset management considers the management of existing equipment and facilities in the transmission network. Transmission and distribution components are expensive, and therefore, they must be utilized in the most efficient way. Since the late 1990s, the power industry has been substantially deregulated and the economic optimization led to an increased interest in asset management. Asset management is closely related to grid development and system operation, as it forms a bridge between these main processes.

A key activity of asset management is maximizing the return on costs of equipment over the entire life cycle by maximizing performance and minimizing investment costs and operational costs for a given risk level [19]. The investment costs (*CAPEX: capital expenditure*) are related to the installation of new assets, while the operational costs (*OPEX: operational expenditure*) are made during the operational lifetime of an asset (like maintenance costs). As the optimization of performance and costs includes the installation of new assets in the power system, asset management is closely related to grid development.

Maintenance forms a crucial part of asset management. Other key activities are: determination of asset condition and lifetime assessment, replacement/upgrade plans, and asset outage management. Asset management can be classified based on the *information domain* and *time domain*. When classified in the information domain, the technical, economical, and societal aspects of asset management are studied [20]. According to the time domain, asset management can be categorized into:

- *Long-term asset management*: The focus is on replacement, refurbishment, or upgrading of existing assets. The time frame usually ranges from a year and beyond.
- *Mid-term asset management*: This involves optimal scheduling of equipment maintenance to extend the life span of existing facilities. The time frame ranges from a few months to a year.
- *Short-term asset management*: Short-term asset management can be subdivided into *operational asset management* (daily/weekly) and *real-time asset management*. Operational asset management aims at minimizing risks involved with assets, both physical and financial, due to loading of the equipment. Real-time asset management is also called *asset outage management*. The analysis of unexpected outages (i.e. contingencies) forms a vital part.

1.4.4 System Operation

System operation considers the operation of the existing power system. Power system reliability is of primary concern in system operation, as it is important to maintain both system security and adequacy at the acceptable levels with minimum socioeconomic cost. System operation encompasses two sublevels, namely *operational planning* and *real-time system operation*:

- *Operational planning*: Operational planning is performed at several instances prior to the establishment of the system operating conditions. It constitutes the preparatory phases before real-time operation. Also, operational planning ensures that the right decisions are taken in advance such that reliable operation is achievable within a prolonged future period of time, called the *operational planning horizon*. The horizon consists of a sequence of target time intervals. The operational planning time horizon can be *week-ahead (W-1)*, two days *(D-2)* in advance, or one *day-ahead (D-1)* as well as several (n) *hour-ahead (H-n)* before real time.
- *Real-time operation*: Real-time operation considers system operation for a time range of about 1 hour to real time. During this time interval, it is assumed that the system operating conditions (like scheduled generation, load, and inter-area exchange) are predictable, although unplanned contingencies and events can always occur. Real-time operation follows a series of activities, planned in a sequential manner. It starts with *preventive control*, which has a time horizon of 1–2 hours, aiming at optimal operation under given security constraints. Furthermore, preventive control oversees possible contingencies and prepares or adjusts the system by taking preventive control decisions like switching equipment and rescheduling generation and load. Preventive control may be followed by two other control strategies, namely corrective control and emergency control, depending on the contingencies and events that occur in the system. *Corrective control* normally considers a time range of 15 min to real time, and it aims at maintaining the system intact when certain system issues (like overvoltages and overloads) are detected. *Emergency control* is performed in real time during unforeseen circumstances (i.e. serious disturbances/outages) following an unplanned contingency or event, and it always aims at maintaining (most of) the system intact at an acceptable security level.

In system operation, it is common practice to use a *framework of operational risk categories* like the one presented in [21]. This framework generally consists of five operating states: *normal, alert, emergency, extreme emergency* and *restorative*. The TSO activities that are performed in system operation are related to the states of this risk framework and to the level of redundancy within the network, as shown in Fig. 1.6. Although essentially the same, the risk framework can vary among different TSOs.

Fig. 1.6 shows that in normal operation, the network is minimally $N - 1$ redundant, such that a single failure will not lead to a failure of the system. In the alert state, there are no serious problems in the network yet, but one more contingency can lead to serious system issues. Because there are just enough components in operation, the alert state can be regarded as the $N - 0$ state. Similarly, the emergency state corresponds to the $\geq N + 1$ state [22], as there are serious overloads and/or overvoltages in the network because too many components failed.

The system states of the risk framework are directly related to the operational control actions. For example, during an emergency, there are serious overloads and/or overvoltages in the network and the TSO has to take action immediately. *Corrective generation redispatch* (a rescheduling of the generation) and *load curtailment*

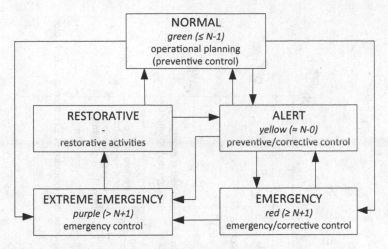

Fig. 1.6 Risk framework, redundancy levels, and TSO activities

(the disconnection of load) are effective actions to relieve the network and to restore $N - 0$ operation. When the network is in the alert state $(N - 0)$, there are no serious problems yet, but the TSO can perform preventive actions to avoid possible emergency situations. *Preventive redispatch* is one of the actions taken in the alert state. The objective of preventive redispatch is to restore $N - 1$ redundancy. During normal operation, the network is $N - 1$ redundant and no corrective actions are required. Still, operational planning is performed, and the TSO can decide to perform preventive control actions when necessary. Considering preventive and corrective control actions, there are in fact two kinds of remedial actions:

1. *Modifying the network:* This can be done by switching actions, influencing the power flow by using phase-shifting transformers (PSTs), or by canceling maintenance activities.
2. *Changing the load/generation:* This can be done by preventive/corrective redispatch, wind curtailment, or load curtailment.

In practice, the different remedial actions have different impacts. Switching actions and the use of PSTs do not influence the generation and load and are therefore generally preferred. If not effective enough, generation redispatch can be applied or maintenance activities could be canceled. Wind and load curtailment are usually considered the last resort.

1.4.5 Overview of TSO Activities

Figure 1.7 gives an overview of the TSO activities, three main processes, time horizons, and time scales. It shows examples of the activities that are performed in each

Main Processes	Long-Term	Mid-Term	Short-Term	Real-Time
Grid Development	Onshore/offshore grid expansion plans (based on generation/load scenarios)	Investments in new grid components (connections, substations, etc.)	Small modifications of the grid (new phase shifters, protection systems, etc.)	---
Asset Management	Refurbishment, replacement and upgrade plans of existing assets	Maintenance scheduling, allocation of resources	Repair and condition monitoring of assets	Condition monitoring, outage management
System Operation	Operational policies	Day-ahead planning	Hour-ahead planning, preventive control actions (redispatch, transport restrictions, canceling maintenance)	Corrective control actions (redispatch, switching actions, wind/load curtailment)

TIME SCALE — Time Horizons

Decades, Years, Months, Weeks, Days, Hours, Minutes

Fig. 1.7 Actions performed by the Transmission System Operator [1]. ©2016 IET, reproduced by permission of the Institution of Engineering & Technology

process and time horizon, while also indicating the actual time scale of these activities. In the figure, the time scales are indicated by diagonal lines, which means that similar time horizons of different main processes consider different time scales. This sometimes causes confusion in practice as, for example, mid-term grid development considers another time scale than mid-term system operation. It is therefore good to be aware of this difference in concept between the three main processes.

There is always an overlap between the different processes, as illustrated in Fig. 1.8. For example, small modifications of the network can be required because of grid development or because of asset management (area 4a in Fig. 1.8). Furthermore, planned maintenance might be canceled during system operation because of a new contingency (area 5a). In the past, the three processes consisted of more or less separate activities. This is illustrated as the *sequential approach* in Fig. 1.8. An example is the Dutch 380 kV ring, which was developed for $N - 2$ ($N - 1$ during maintenance) redundancy (area 1a). This gave enough room to plan maintenance in asset management (area 2a) and enough room for operational activities (area 3a). In system operation, actions were performed in the network 'as is.'

In the future, the overlapping and interaction between the three main processes is expected to increase because of further optimization of the usage of the power system infrastructure. Figure 1.8 illustrates this as the *interacted approach*. An illustrative example is the development of offshore grids. Previous studies have shown that offshore network redundancy is mostly not economical. However, to maintain a high level of security of supply, redundancy must be created differently. For example, onshore spinning generation reserve can serve as redundancy for the offshore network (area 6b in Fig. 1.8). If redundancy is not created in the offshore network (a long-term

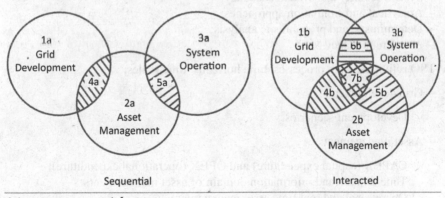

Sequential   Interacted

1. Investments on new infrastructure
2. Condition monitoring, maintenance
3. Operational planning, real-time operation
4. Small modifications of the network: replacement, refurbishment, upgrades
5. Maintenance planning/canceling
6. Grid investments vs operational control actions
7. Integrated network design (grid development, asset management and operational planning)

Fig. 1.8 Interactions among the three processes [1]. ©2016 IET, reproduced by permission of the Institution of Engineering & Technology

grid development activity), this has consequences for activities in other time horizons. For example, operational policies need to be adapted, and generation redispatch or load curtailment is more often needed in short-term system operation.

Problems

1.1 Learning Objectives
Please check the learning objectives as given at the beginning of this chapter.

1.2 Main Concepts
Please check whether you understand the following concepts:

- Organization of the power system:

  - Transmission and Distribution System Operator (TSO and DSO);
  - EHV, HV, MV, LV (extra high, high, medium, and low voltages);
  - Liberalization of the electricity market;
  - Consumers, producers, system operators, customers.

- Reliability and risk:

  - System and component;
  - System reliability;
  - Risk;
  - Risk matrix and business values;
  - System security and system adequacy;
  - Analytical and simulation approaches;
  - Deterministic and probabilistic analysis;
  - $N - 1$ redundancy.

- TSO activities, main processes, time horizons, time scales:

  - Grid development:

    · Development scenarios.

  - Asset management:

    · CAPEX (capital expenditure) and OPEX (operational expenditure);
    · Time domain and information domain of asset management;
    · Operational and real-time asset management.

  - System operation:

    · Real-time operation and operational planning;
    · Week-ahead (W-1), day-ahead (D-1), n hour-ahead (H-n);
    · Corrective, preventive, and emergency control;

· Risk framework (normal, alert, emergency, extreme emergency, restorative);
· Corrective/preventive generation redispatch, load curtailment.

– Sequential and interacted approach of TSO processes

1.3 System Reliability
What is the basic definition of system reliability? And what is (or are) the main function(s) of the power system?

1.4 Liberalization Electricity Market
Describe how the liberalization of the electricity market influenced the concept of power system reliability.

1.5 Business Values
Mention some business values used in the risk matrix of TenneT, and describe how they reflect the reliability of the transmission network.

1.6 Probabilistic & Deterministic Analysis
What are the main differences between deterministic and probabilistic power system reliability analysis?

1.7 TSO Activities
Mention some TSO activities in the long term, mid term, and short term for the three processes—grid development, asset management, and system operation.

1.8 Interaction TSO Activities
Give an example of how grid development is related to system operation, seen from a network reliability point of view. Also give examples of the interaction between grid development and asset management, and between asset management and system operation. How will the overlap between the main processes change in the future, and why?

1.9 Control Actions
Mention some corrective and preventive control actions. In which order are they preferred?

References

1. Khuntia SR, Tuinema BW, Rueda JL, van der Meijden MAMM (2015) Time horizons in the planning and operation of transmission networks: an overview. IET Gener, Transm Distrib 10(4):841–848
2. Klaassen KB, van Peppen JCL, Bossche A (1988) Bedrijfszekerheid, Theorie en Techniek. Delftse Uitgevers Maatschappij, Delft, The Netherlands
3. Billinton R, Allan RN (1996) Reliability evaluation of power systems, 2nd edn. Plenum Press, New York
4. Cepin M (2011) Assessment of power system reliability—methods and applications. Springer, London

5. Autoriteit Consument en Markt (2014) Netcode Elektriciteit. Autoriteit Consument en Markt, The Netherlands
6. Li W (2005) Risk assessment of power systems—models, methods, and applications. Wiley Interscience - IEEE Press, Canada
7. TenneT TSO B.V. (2018) TenneT asset risk matrix, update 2018. TenneT TSO, Arnhem, The Netherlands
8. Hass D, Pels Leusden G, Schwarz J, Zimmermann H (1981) Das (n-1)-Kriterium in der Planung von Übertragungsnetzen. Elektrizitätswirtschaft 80(25)
9. Cigré Working Group C4.601 (2010) Review of the current status of tools and techniques for risk-based and probabilistic planning in power systems. Cigré, Paris
10. Siemens (2011) PSS(R)E product description. http://www.energy.siemens.com
11. DIgSILENT GmbH (2011) PowerFactory product description. http://www.digsilent.de
12. Zimmerman RD, Murillo-Sanchez CE, Thomas RJ (2011) Matpower: steady-state operations, planning and analysis tools for power system research and education. IEEE Trans Power Syst 26:12–19
13. Bertling L, Bangalore P, Tuan LA (2011) On the use of reliability test systems: a literature survey. In: IEEE Power and Energy Society General Meeting (PES-GM2011), Detroit, MI, USA, pp 1–9
14. Probabilistic Methods Subcommittee (1979) IEEE reliability test system. IEEE Trans Power Appar Syst 98:2047–2054
15. Billinton R, Kumar S, Chowdhury N, Chu K, Depnath K, Goel L, Khan E, Kos P, Nourbakhsh G, Oteng-Adjei J (1989) A reliability test system for educational purposes—basic data. IEEE Trans Power Syst 4:1238–1244
16. Billinton R, Kumar S, Chowdhury N, Chu K, Depnath K, Goel L, Khan E, Kos P, Nourbakhsh G, Oteng-Adjei J (1990) A reliability test system for educational purposes—basic results. IEEE Trans Power Syst 5:319–325
17. Wood AJ, Wollenberg BF (2012) Power generation, operation, and control. Wiley, Hoboken
18. TenneT TSOBV (2008) Visie2030. TenneT TSO B.V., Arnhem, The Netherlands
19. Tor O, Shahidehpour M (2006) Power distribution asset management. In: IEEE power engineering society general meeting, 2006. PES-GM'06
20. Smit JJ, Quak B, Gulski E (2006) Integral decision support for asset management of electrical infrastructures. In: IEEE systems, man and cybernetics conference
21. Billinton R, Fotuhi-Riruzabad M (1994) A basic framework for generation system operating health analysis. IEEE Trans Power Syst 9(3):1610–1617
22. Schilling MT, Rei AM, do Coutto Filho MB, Stacchini de Souza JC, (2002) On the implicit probabilistic risk embedded in the deterministic n-α type criteria. In: 7th international conference on probabilistic methods applied to power systems (PMAPS2002), Napels, Italy

Chapter 2
Power System Failures

Power systems can fail because of various reasons. Often, the effect of smaller failures is limited to the failure and repair of a single network component, as the power system is designed and operated considering $N - 1$ redundancy. Serious failures of the power system are therefore caused by combinations of component failures and other influencing factors. In probabilistic reliability analysis, the reliability of the power system is determined based on the reliability of individual components. Failure statistics of components are the base of the reliability analysis and can be derived from databases with historical component failures. Section 2.1 discusses some typical failure statistics of power system components. These failure statistics can be used to create reliability models of power system components. Specific knowledge about the failure behavior of components is useful as well, as discussed in more detail in Sect. 2.2. Historical blackouts can provide valuable information about how power systems can possibly fail. By analyzing these blackouts, it can be investigated which series of events led to these large blackouts. As will become clear from the examples discussed in Sect. 2.3, large blackouts are mostly caused by multiple component failures, where dependency and other influencing factors play an important role. Based on this knowledge, it can then be decided which factors should be taken into account in reliability analysis to calculate more realistic results. Suggestions to improve probabilistic reliability analysis in the future by including more of these influencing factors are discussed in the conclusions of this chapter in Sect. 2.4.

Learning Objectives

After reading this chapter, the student should be able to:

- explain what failure statistics are and how they are collected and presented;
- understand how components fail (causes, dependency/correlation, effects);
- explain how individual component failures can lead to large blackouts;
- discuss the challenges of (probabilistic) power system reliability analysis.

© Springer Nature Switzerland AG 2020
B. W. Tuinema et al., *Probabilistic Reliability Analysis of Power Systems*,
https://doi.org/10.1007/978-3-030-43498-4_2

2.1   Component Failure Statistics

Failure statistics, i.e. information about historical failures or test results of network components, are the main input information for reliability analysis. An accurate description of these statistics will lead to more accurate results. It can, however, be challenging to collect suitable failure statistics for particular components. Failure statistics can be scarce, as some technologies are relatively new and most of the components are designed to be reliable. To obtain more failure statistics, either the *observed population* or the *observation period* can be increased, but both solutions have their complications. When, for example, extra-high-voltage (EHV) underground cables (UGCs) are considered, the amount of failure statistics from one power system might not be sufficient. In this case, the observation period cannot be increased because EHV cables are relatively new technology. Because of the different technology, a combination with failure statistics of high-voltage (HV) cables is not possible either. To increase the accuracy, the observed population can be increased by combining failure statistics with those of other EHV power systems. When combining failure statistics of different power systems, the characteristics and circumstances must be comparable. For example, failure statistics of the Dutch and German power systems could be compared and combined. By contrast, failure statistics of the Dutch power system probably cannot be combined with statistics from the Spanish power system because of the different environments.

In the Netherlands, failures of power system components are recorded in the NESTOR (NEderlandse STOringsRegistratie) database [1]. For each failure, information like the component, the failure cause and the repair time is reported. In NESTOR, failures of the extra-high-voltage (EHV) network (220/380 kV) have been reported since 2006. Data for failures in the high-voltage network (110/150 kV) is available for 1998 and later. Extracts from NESTOR are also published in yearly reports called 'Betrouwbaarheid E-Netten' [2–10]. In Germany, failures of power system components are stored in a database by VDN and published in reports as well [11, 12]. The following sections discuss some typical values of failure frequencies and repair times that can be derived from these databases.

2.1.1   Failure Frequencies

From the NESTOR database, it is possible to calculate average failure frequencies of specific network components. The precise definition and calculation of the failure frequency will be explained in detail in Chap. 3, but for now, it suffices to know that the failure frequency shows how often a component fails per year, on average. The graphs in Fig. 2.1 show the average failure frequencies together with their confidence intervals for the study period from 2006 up to (and including) 2013. The confidence intervals are an indication of the accuracy and depend on the amount of available failure statistics. The figures show that overhead lines, underground cables, trans-

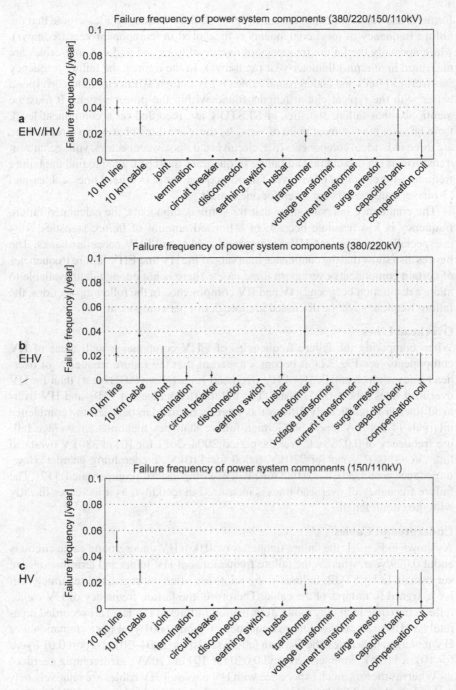

Fig. 2.1 Failure frequency of various power system components (NESTOR 2006–2013). The dots represent the averages (Appendix A), and the lines indicate the (χ^2-95%) confidence interval

formers, and capacitor banks have the highest failure frequency. Please note that the failure frequency of most components is measured in /component-year (/compyr), while normally the failure frequency of overhead lines and underground cables are measured in /circuit-kilometer-year (/cctkmyr). In the figures, the failure frequency for overhead lines and underground cables is given for a 10 km circuit to correspond better with the typical circuit lengths found within the power system. It must be mentioned that failure statistics in NESTOR are recorded on a component level, from an asset management point of view. As there are three phases in a circuit, there are three individual components (e.g. circuit breakers, disconnectors, voltage/current transformers) in one circuit and failure frequencies must be tripled to find the failure frequency for one circuit. However, failure frequencies of overhead lines, cables, and (3-phase) transformers are given for one circuit.

The confidence intervals show that for some components, the calculated failure frequency is less accurate because of a limited amount of failure statistics. This is especially the case for EHV underground cables and EHV capacitor banks. The figures also show that the confidence intervals of the HV and EHV failure frequencies of certain components overlap. In these cases, there is not enough data available to make a distinction between EHV and HV components. In the following sections, the failure frequency will be discussed in more detail per component.

Overhead Lines

When comparing the failure frequencies of EHV components with these of HV components (see Fig. 2.1), it becomes apparent that the failure frequency of overhead lines is much smaller for EHV overhead lines (about 0.021/year) than for HV overhead lines (about 0.051/year). The failure frequencies of EHV and HV overhead lines are significantly different as there is no overlap between the confidence intervals of both. A report on German failure statistics mentions an average failure frequency of 0.026/year over the period 2004–2011 for 10 km 380 kV overhead line circuit, 0.037/year for 220 kV, 0.060 for 110 kV 'niederohmig geerdet' (low-impedance grounded), and 0.022 for 110 kV 'kompensiert' (compensated) [12]. The failure frequency of overhead lines is measured in /cctkmyr, as it increases linearly with the circuit length.

Underground Cables

As shown in Fig. 2.1, the failure frequency of 10 km HV underground cable circuit is about 0.029/year, whereas the failure frequencies of HV joints and terminations are very small. In NESTOR, failures of the joints are often not registered separately, but are assigned to failures of the cable. Therefore, the failure frequency of HV cables reflects the cable including joints. Termination failures at HV level are recorded separately and have an average failure frequency of about 0.001/year per termination. For HV underground cables, the German failure statistics (2004–2011) report 0.014/year for 10 km 110 kV 'kompensiert' and 0.050 for 10 km 110 kV 'niederohmig geerdet.'

Whereas there is much experience with HV cables, EHV cables are relatively new technology. In the Dutch and German power systems, only a few short 380 kV cables are installed and are mostly used to connect individual power plants to a substation. As the amount of available failure data of EHV underground cables is very small,

Table 2.1 Cable system failure frequencies from different sources

Component	Cigré379 [13]		TSOs high [14]		TSOs low [14]	
Cable	0.00133[a]	[/cctkm·y]	0.00120[a]	[/cctkm·y]	0.00079[a]	[/cctkm·y]
Joint	0.00048[b]	[/comp·y]	0.00035[b]	[/comp·y]	0.00016[b]	[/comp·y]
Termination	0.00050[b]	[/comp·y]	0.00168[b]	[/comp·y]	0.00092[b]	[/comp·y]

[a]The failure frequency of the cable measured in [/circuit-kilometer-year] is the failure frequency of a single cable circuit with one individual cable per circuit phase
[b]The failure frequency of the joints and terminations measured in [/component-year] is the failure frequency of individual components. For a 3-phase circuit consisting of one individual cable per circuit phase, sets of three joints (and terminations) are required

Fig. 2.1 shows a large confidence interval for EHV underground cables. Currently, in the Netherlands 20 km (double-circuit) underground cable has been installed in the 380 kV transmission network in the Randstad380 project. The total cable length and the fact that the underground cables are part of the backbone of the EHV transmission network make this project rather unique. Because of the limited amount of failure data for EHV cables, the failure frequency cannot be based on the NESTOR database, nor can it be based on the German reports alone.

In Cigré Technical Brochure 379, failures of underground cables all over the world were reported [13]. Still, the number of cable failures at higher voltage levels (380–500 kV) was limited. In a survey among European TSOs [14], failures of EHV cables were studied specifically. High and low estimates of the failure frequencies of cable circuit components were calculated and are shown in Table 2.1. The high estimate probably is more accurate as it is based on failure data of only these particular TSOs that responded to the survey. For the low estimate, all TSOs were included and it was assumed that the TSOs that did not report any failures did not experience any cable failures.

Transformers

For transformers, Fig. 2.1 shows a difference in the failure frequencies of EHV and HV (high-voltage side of the transformer) transformers. The failure frequency of EHV transformers is about 0.039/year and is 0.014/year for HV transformers. The German failure statistics (2004–2011) show a failure frequency of about 0.024/year for 380 kV transformers, 0.028/year for 220 kV transformers, 0.015/year for 110 kV 'niederohmige Erdung,' and 0.009/year for 110 kV 'kompensiert' [12]. From the studied data sources, it is difficult to explain the difference. A difference does not necessarily mean that there is a difference in the quality of the components. Other failure registration techniques may cause a difference as well. Also, variations of the average failure frequency in time are possible. The German statistics report an average failure frequency of 0.007/year for 110 kV 'niederohmige Erdung' for the period 1994–2001, while the other failure frequencies are comparable with the values for the period 2004–2011.

Substation Equipment

As shown in Fig. 2.1, the failure frequencies of substation equipment (busbars, measurement transformers, breakers/switches) are very small. For HV substation equipment, Cigré working group A3.06 studied the reliability of circuit breakers, disconnectors/earthing switches, instrument transformers and gas-insulated switchgear (GIS) [15].

For busbars, NESTOR reports a failure frequency of 0.002/year. For circuit breakers, a Cigré report showed a failure frequency of 0.003/year [15]. The failure frequency of disconnectors and earthing switches, as mentioned by Cigré, is 0.003/year. For instrument transformers (i.e. current transformers and voltage transformers), Cigré reports a failure frequency of about 0.0001/year. NESTOR reports a failure frequency of 0.018/year for capacitor banks, 0.003/year for compensator coils, and 0.001/year for surge arrestors. It must, however, be mentioned that due to the limited amount of data, the confidence interval of these values is large.

Overview of Failure Frequencies

Table 2.2 gives an overview of the failure statistics discussed in this section.

2.1.2 Repair Times

Average Repair Times

Table 2.2 Overview of the failure frequency of power system components

Component	Failure frequency	Source
Overhead line (EHV)	0.0021 (/cctkmyr)	NESTOR [1]
Overhead line (HV)	0.0051 (/cctkmyr)	NESTOR [1]
Cable (HV) (including joints)	0.0029 (/cctkmyr)	NESTOR [1]
Termination (HV)	0.001 (/compyr)	NESTOR [1]
Cable (EHV)	0.00120 (/cctkmyr)	European TSOs [14]
Joint (EHV)	0.00035 (/compyr)	European TSOs [14]
Termination (EHV)	0.00168 (/compyr)	European TSOs [14]
Transformer (EHV)	0.039 (/compyr)	NESTOR [1]
Transformer (HV)	0.014 (/compyr)	NESTOR [1]
Busbar (EHV/HV)	0.002 (/compyr)	NESTOR [1]
Circuit breaker (EHV/HV)	0.003 (/compyr)	Cigré [15]
Disconnector/earthing switch (EHV/HV)	0.003 (/compyr)	Cigré [15]
Instrument transformer (EHV/HV)	0.0001 (/compyr)	Cigré [15]
Surge arrestor (EHV/HV)	0.001 (/compyr)	NESTOR [1]
Capacitor bank (EHV/HV)	0.018 (/compyr)	NESTOR [1]
Compensation coil (EHV/HV)	0.003 (/compyr)	NESTOR [1]

Table 2.3 Average repair times, 3rd-quartile values from boxplots, and repair times according to the exponential approximation

Component	T_{av} (h) (average)	3rd-quartile (h) (boxplot)	T_0 (h) (exponential)
Overhead line (EHV/HV)	44	8	4
Cable (HV)	317	249	146
Joint/termination (HV)	607	248	146
Circuit breaker (EHV/HV)	409	59	19
Disconnector/earthing switch (EHV/HV)	578	248	6
Voltage/current transformer (EHV/HV)	735	169	28
Transformer (EHV/HV)	329	23	5
Busbar (EHV/HV)	318	3	2
Capacitor bank	1587	140	14
Surge arrestor (EHV/HV)	144	2	2

Repair times are, together with the failure frequencies, the most important input parameters for reliability analysis. From the failure statistics reported in NESTOR, the average repair times of the discussed components can be calculated. The results are shown in Table 2.3 (second column). As can be seen, these average repair times are rather long. When calculating average repair times, it is often found that several extremely long repair times dominate the calculated average. This effect can be illustrated by boxplots.

Boxplots of Repair Times

A *boxplot* is a graphical way to present statistical data. It shows the *minimum* value of the population (or data), the 1st-*quartile* (25% of the population is smaller than this value), the *median* (50% of the population is smaller than this value), the 3rd-*quartile* (75% of the population is smaller than this value), and the *maximum*. *Outliers* are sometimes excluded from the boxplot. Figure 2.2 shows an example boxplot of the repair times of overhead lines. It can be seen that 75% of the overhead line failures were repaired within 8 h.

Figure 2.3 shows the boxplots of the repair times of various components, based on the NESTOR database [1]. It can be seen that there is a large variation in the repair time of power system components. For example, failures of overhead lines are mostly repaired within hours, whereas the repair of cable circuits (i.e. cable, joints, and terminations) can take hundreds of hours. The boxplots further show that disconnector/earthing switches, instrument transformers, and capacitor banks have a relatively long repair time. The repair times of circuit breakers, transformers, and busbars are relatively short. The short repair time of circuit breakers and transformers might seem a bit illogical because one might expect that a repair or replacement of these complicated components will take much longer. In the database, however, a number of circuit breaker and transformer failures were repaired within a short amount of time. Probably, in most of these cases, no complicated repair or replace-

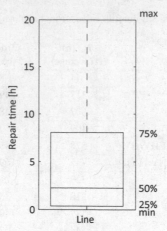

Fig. 2.2 Boxplot example for repair times of overhead lines

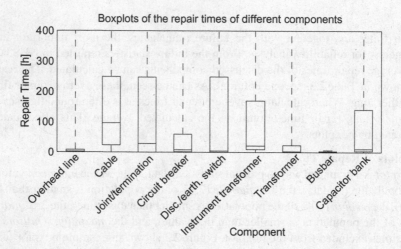

Fig. 2.3 Boxplots (Appendix A) of the repair times of different system components

ment was needed. In general, every component can have a short repair time (if the component has not been seriously damaged) or an extremely long repair time (if the component needs to be replaced).

As comparison, the 3rd-quartile values of the boxplots are shown in Table 2.3 as well. It can be seen that these values are often much smaller than the average repair time, meaning that (much) more than 75% of the component failures are repaired within the average repair time.

Exponential Approximations and Averages

Although boxplots are very useful for visualizing data, boxplots are less useful for further reliability analysis in practice as it is not clear which number (e.g. median,

3rd-quartile, or average) should be used in further calculations. As will be explained later in Chap. 3, normally the repair process is approximated by the (negative) exponential distribution. The repair time can then be estimated accurately by fitting the repair time data to this exponential distribution. Table 2.3 shows the repair times according to the exponential distribution as well. These values are smaller than the 3rd-quartile boxplot values. Often, the average repair time is strongly influenced by several extremely long repair times. This effect is canceled out when performing a curve fit to the repair time data, while extremely long repairs are still possible in the exponential distribution. In this sense, simply using the average repair time in reliability calculations could lead to results that are too pessimistic.

For probabilistic reliability analysis, it is often challenging to find accurate, well-defined repair times. Because repair times can vary much between different data sets, *expert's opinion* about the expected average repair times can be an alternative way to estimate repair times when sufficient data is lacking.

EHV Underground Cables

For underground cables, the repair times shown in Table 2.3 are for HV cable circuits only. Because of the relatively new technology, the repair time of EHV underground cables is expected to be significantly longer. In a Cigré report, a repair time of 600 h for EHV cable circuits is mentioned [13]. However, a survey of EHV underground cable failures among European TSOs shows repair times ranging from 2 weeks to 2 months, and even one extremely long repair time of 3/4 year [14]. Therefore, it is advisable to take a longer repair time into account for EHV cable circuits (consisting of cables, joints, and terminations). As the number of failure statistics of EHV cables is very limited and the extremely long repair times were mainly caused by logistical issues, an estimation of the repair time of EHV cables is made based on expert's opinion. It is assumed that the current expected repair time is 730 h (1 month), which can be reduced to 336 h (2 weeks) in the future by using a more optimized repair scheme [16].

2.2 Component Failure Behavior

To study the reliability of power systems, not only the failure statistics of components must be known, but also the failure behavior of components must be understood. In this section, it is explained how components of the power system can fail. The causes as well as the effects are discussed and concepts like (in)dependent failures and correlation are introduced.

2.2.1 Concepts and Definitions

First of all, some basic concepts and definitions related to component failures need to be discussed. A *failure* of a component is the termination of its ability to perform a required function [17, 18]. The cause that initiated the failure is called the *failure cause*, while the physical process that leads to the failure is called the *failure mechanism* [19, 20]. The observed appearance of the failure is then called the *failure mode* [19, 20]. A component failure might result in a (system) *fault*. According to [17, 21], a fault is an abnormal condition that may cause a reduction in, or loss of, the capability of a functional unit to perform a required function. In this case, the functional unit is the power system. According to [22], a fault in a circuit is any failure which interferes with the normal flow of current. Combining these two definitions, a fault in a power system can be defined as an abnormal electrical condition which interferes with the normal power flow and may affect the capability of the power system to fulfill its function.

As an example, a possible failure cause in underground cables is an impurity in the insulation. This impurity can initiate the failure mechanism treeing, which leads to an insulation breakdown as failure mode, and eventually, this can cause a line-ground fault in the cable circuit. Component failures do not always lead to a fault. For example, if a transformer is switched off because of oil leakage, it is a component failure without a fault. Sometimes, there is a fault without a component failure. For example, when there is a flash-over between an overhead line circuit phase and a tree branch, the phase is disconnected by the protection system and almost immediately reconnected by *auto-reclosure*. Although component failures are often detected immediately, some component failures remain undiscovered for a while. An example of such a *hidden failure* is a stuck disconnector, which is only discovered when the operation is demanded.

When a component is switched off, this causes an *outage* in the power system. A distinction can be made between *forced outages* and *planned outages* [23]. *Forced outages* are caused by failures of components switched off immediately by the protection system, but also sometimes disconnected manually by the system operator immediately. *Planned outages* are maintenance activities, which can be initiated when a component type is known to be vulnerable to certain failure causes or failure mechanisms. Sometimes, a third group of outages is defined, the *semi-forced outages* [23]. These are cases in which a component still works, but maintenance is required within a short time. Figure 2.4 illustrates the progress from failure cause to an outage. Possible outages within the power system are also called *contingencies*. Then, in *contingency analysis* the power system is analyzed with certain components out of service.

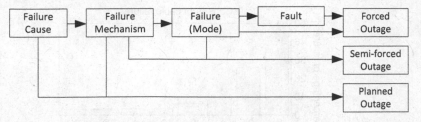

Fig. 2.4 Progress from failure cause to outage

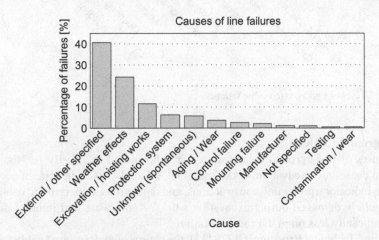

Fig. 2.5 Causes of overhead line failures

2.2.2 Causes of Component Failures

Power system components can fail because of various reasons. The failure statistics from the NESTOR database can provide more information about this [1]. In the following sections, the failure causes of power system components are discussed in more detail.

Overhead Lines

Figure 2.5 shows the causes of overhead line failures. As can be seen, the major part of the failures was caused by external and other, specified, causes (like birds, trees, or a kite hitting a line). Weather effects (like lightning and storms) caused about a quarter of the failures. Hoisting works were the cause of failure in about 10% of the cases. The figure also shows that aging/wear only caused a small amount of the overhead line failures. It can also be seen that a small part of the failures was caused by the protection system, and in some other cases, the real cause remained unknown, even after investigation.

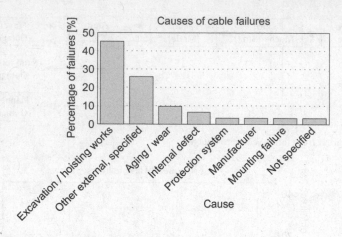

Fig. 2.6 Causes of underground cable failures

Underground Cables

The causes of underground cable failures are shown in Fig. 2.6. By far, the most frequent causes are excavation works and other external causes. In spite of accurate location indicators, protection constructions, and regulations, underground cables are still regularly damaged during excavation works. Aging/wear and internal defects are significantly less often the cause of failure.

In Cigré Technical Brochure TB379 [13], the causes of underground cable failures are mentioned as well: 45% internal failures, 39% third-party mechanical, 13% other physical external parameters (e.g. increased burial depth), and 3% abnormal system conditions (e.g. lightning) for 60–500 kV direct burial cables. Compared to NESTOR, Cigré reports less external and more internal causes.

Circuit Breakers

In a survey, Cigré studied the service conditions of failed circuit breakers [15]. The results show that 45% of the circuit breakers was in normal service and operation was demanded, 30% was in normal service without operation command, 8% was de-energized and available for service, 7% of the failures occurred during or directly after testing/maintenance, and 6% occurred during fault clearing. The main failure modes were not closing/opening on command (37%) and locked in open/closed position by the control system (30%). This means that most circuit breaker failures are operational failures, which are likely to be discovered only when the operation is demanded (i.e. hidden failures). It seems therefore more logical to define circuit breaker failures in failures per switching action in probabilistic reliability analysis, although from an asset management and maintenance perspective it can be useful to know the amount of failures per year. In the reports of German failure statistics [11, 12], operational failures of circuit breakers are defined as a conditional probability, as will be discussed further in Sect. 2.2.3.

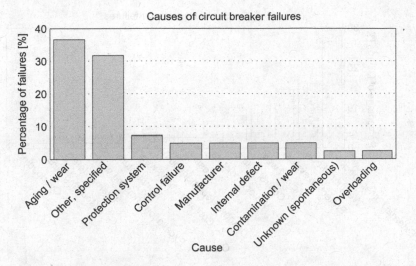

Fig. 2.7 Causes of circuit breaker failures

Most of the circuit breaker failures reported in NESTOR were caused by aging/wear, as shown in Fig. 2.7. Cigré states that 40% of the circuit breaker failures are caused by aging/wear. This indeed is the most logical failure cause. Other causes are diverse (e.g. internal defect, manufacturer, contamination, improper operation).

Disconnectors
Similar to circuit breakers, failures of disconnectors, and earthing switches are mostly operational failures. Cigré shows that in 72% of the failures, the device did not operate on command. The main cause of the major failures was aging/wear (60%). Similar to circuit breakers, it seems more logical to define the failure frequency of disconnectors and earthing switches in failures per switching action (or a conditional probability) in probabilistic reliability analysis, although for asset management it can still be useful to know the failure frequency in failures/year.

Transformers
The causes of transformer failures as reported in NESTOR are shown in Fig. 2.8. External causes and weather effects were the most common causes of transformer failures. Furthermore, the protection system was mentioned in about 15% of the failures. This can refer to a failure of the protection system in some cases, but in other cases, the protection system worked properly to protect the transformer from a failure of a secondary installation. Aging and wear are reported in about 10% of the cases. Other causes were less common.

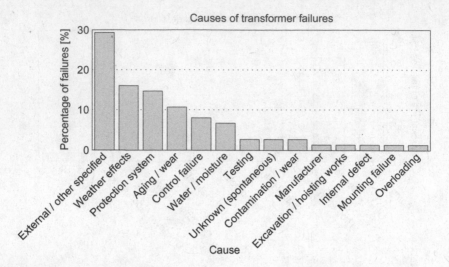

Fig. 2.8 Causes of transformer failures

2.2.3 Protection System Failures

As could already be seen from Figs. 2.5 to 2.8, protection systems are a possible cause of component failures as well. For example, it is possible that a protection system does not clear a system fault properly, and more components are switched off than necessary. It is also possible that a component is switched off by the protection system without any reason (i.e. spontaneous switching). In this section, failures of the protection system are studied in more detail.

In some of the component failures reported in NESTOR, the protection system is mentioned as the cause of failure. In these cases, a further specification describes the actual failure mode of the protection system. Based on the Dutch failure statistics in NESTOR, the overview shown in Table 2.4 was made. The table shows that 12% of the component failures were caused by the protection system (for all voltage levels). Some possible failures of protection systems are spontaneous switching or an improper setting of the protection system. In other cases, there was no switching command or the switching command was too late.

Table 2.4 also shows how protection systems play a role in component failures not caused by the protection system (like weather effects, external causes, and aging/wear). In 9% of the cases, the failed component was not switched off correctively/selectively by the protection system. This value seems rather large, especially when compared with the percentage of cases where the failed component was switched off correctively/selectively (43%). This rather large value is caused by the fact that in the power system, not all components are protected individually. Whereas in higher voltage networks all components are protected individually, in lower voltage networks components are protected in groups because of protection possibilities

Table 2.4 Protection system failures (380/220/150/110 kV)

Caused by the protection system	
Spontaneous switching	2%
Improper setting	2%
No switching command	0.4%
Switching command too late	0.4%
Other, specified	3%
(blank)	3%
Subtotal	**10%**
Caused by unwanted operation of the protection	**2%**
Not caused by the protection system	
Switched off by protection (corrective/selective)	43%
Switched off by hand	20%
Switched off by protection (not corrective/selective)	9%
Not switched off	4%
Not applicable	1%
(blank)	11%
Subtotal	**88%**
Grand total	**100%**

and economics. Furthermore, most of these failure cases are failures of a secondary installation that led to the disconnection of connected components.

As the table shows, an amount of failures was switched off by hand. This does not mean that the protection system did not function properly. In most of these cases, the component failure did not result in a system fault, although the component had to be switched off by hand for repair. In most of the cases where the component was not switched off at all, the failure did not result in a system fault, and probably, there was no need to switch off the component by hand immediately.

Basically, a protection system consists of detection (e.g. by instrument transformers), control, and switching (by circuit breakers). Failures of protection systems are therefore caused by failures of these specific subsystems and for reliability analysis, it would be useful to know the failure probabilities of these subsystems. There are however some issues with the overview of Table 2.4. Although a specification is given for most protection system failures, still in 3% of the cases the specification is not given ('blank'). Another 3% is described as 'other, specified,' while further specification is often missing in the report. Also, the distinction between 'caused by protection system' and 'caused by unwanted operation of the protection' is not clear. It is therefore hardly possible to translate these statistics into failure probabilities of protection systems.

The report on German failure statistics gives values for different protection failure modes [11]. In this report, the conditional probability of a protection system failure after a component failure is 0.001. The probability of a circuit breaker failure is

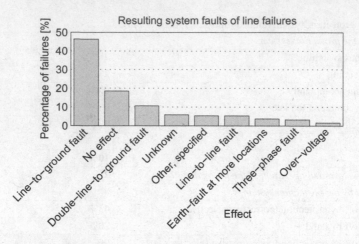

Fig. 2.9 Resulting faults of overhead line failures

about 0.0015, and the probability of unselective switching after a component failure is about 0.005. The spontaneous switching failure frequency is about 0.01/year. The German report does not further describe the origin of these numbers, though.

2.2.4 Resulting Faults

Component failures can lead to various faults. For example, Fig. 2.9 shows the faults that were caused by overhead line failures. In most cases, the failure resulted in a line-to-ground fault (e.g. a tree branch hitting one circuit phase). In almost 20% of the cases, there was not any fault (this can be the case when a circuit is disconnected manually for safety reasons). Other faults like double-line-to-ground, line-to-line, and three-phase faults occurred less often. The study of *symmetrical faults* (i.e. three-phase faults) and *unsymmetrical faults* (like line-to-ground and line-to-line faults) is very important in power system analysis [22].

2.2.5 Dependency and Correlation

The previous sections discussed the failure of individual components. It is also possible that multiple components are involved in one failure event. These *dependent failures* can have large consequences for the network reliability. Two kinds of dependent failures can be distinguished. In *common-cause failures*, there is one cause which directly leads to the failure of multiple components (e.g. a collapsed tower leads to the failure of both electrical circuits carried by the tower). In *cascading failures*, there is one failure first (*primary failure*), which then leads to a *secondary failure* (e.g. one circuit fails, then another circuit fails because of overloading). One can easily imagine that cascading failures must be limited to as few components as possible to prevent large blackouts.

Dependency is related to correlation. Whereas for dependent failures there is a causal relationship between the failures, for correlation a causal relationship is not necessary. For example, the number of overhead line failures might be correlated with the number of cars on the road. There is, however, no dependency because the number of overhead line failures is not the result of the number of cars on the road. Correlation in this case only means that there are more cars on the road during daytime and more overhead line failures during the daytime. *Correlation* therefore is a relation based on common behavior or patterns, and dependency is a subgroup of correlation. The study of dependent and correlated failures provides information about whether multiple failures occur under certain circumstances. Probabilistic reliability analysis can be made more realistic by including this in the analysis.

Dependent Failures

Failure statistics can provide information about the occurrence of dependent failures. There are, however, some challenges. Firstly, failure statistics of some components are scarce, and dependent failures of these components might not have occurred yet. Secondly, it is not always clear which failures in a database are dependent and which are not.

In the NESTOR database, 20 independent (single-circuit) failures as well as 2 dependent double-circuit failures were reported for EHV (220/380 kV) overhead lines. The amount of dependent double-circuit failures is thus 10% the amount of independent (single-circuit) failures. The German reports show a failure frequency of dependent double-circuit failures of 10% the failure frequency of independent failures [11, 12]. Based on this limited information, it is a reasonable assumption that the failure frequency of dependent double-circuit overhead line failures is about 10% the failure frequency of independent (single-circuit) overhead line failures.

Table 2.5 gives an overview of the amount of involved components per failure, as reported in NESTOR. For this overview, it was assumed that failures that occurred within half an hour within the same area belong to the same failure event. It can be seen that in 92% of the cases, only one component was involved. In 8% of the

Table 2.5 Amount of involved components per failure

Amount of components	Percentage of failures (%)
One	92
Two	5
Three	2
Four	1
Six	0.2
Eleven	0.2

Table 2.6 Amount of involved line circuits per failure

Amount of components	Percentage of failures (%)
One (independent)	84
Line circuit + component	5
Double circuit	4
Two line circuits	2
Three line circuits	2
Four line circuits	1
Five line circuits	1
Six line circuits	1

cases, more than one component failed. In these cases, up to eleven components were involved in the same failure event. In the last case, multiple overhead lines were switched off as a result of stability problems. A similar overview can be made for the amount of overhead line circuits involved in the same failure event. The result is shown in Table 2.6. It can be seen that in 84% of the failures, only one line circuit was involved, while in 16% of the cases more circuits were involved.

These studies show the significance of dependent failures. As will become clear when discussing historical blackouts in Sect. 2.3, large blackouts are often caused by dependent failures and combinations of events. Therefore, dependent failures must be included in reliability analysis to obtain realistic results. Although based on limited information, the discussed studies all showed a percentage of dependent failures of roughly 10%. A dependent failure factor of 10% therefore seems a reasonable assumption for further reliability studies, like the ones described in the following chapters of this book.

Correlation

Component failures can be correlated to other variables, like system load or time. As already mentioned before, correlation does not necessarily mean that there is a causal relationship. For some variables, like weather effects, a correlation of the failure frequency and the month of the year can be expected. For example, severe storms are more likely in fall, whereas lightning is more common in summer. If a correlation is strong enough, it can be included in probabilistic reliability analysis to make it more realistic. In this section, the correlation between failure statistics from NESTOR and time is analyzed.

All failures recorded in NESTOR can be arranged to the month of the year, which leads to the top graph of Fig. 2.10. As can be seen, there is no clear correlation between the amount of failures and the month of the year. Of course, this is an overview of all failures, caused by all possible causes. It may therefore be better to consider each failure cause separately to study this kind of correlation. If this is done for the NESTOR data, no significant correlation with the month of the year can be found for any failure cause. For weather effects, this can be clarified by the fact that

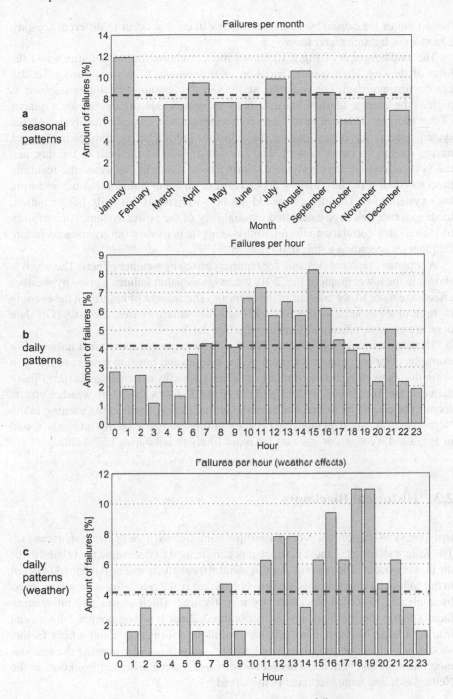

Fig. 2.10 Seasonal and daily patterns in the number of component failures

these failures are caused by various weather effects that occur in different seasons, like storms, lightning, and snow.

The middle graph of Fig. 2.10 shows the correlation between failures and the hour of the day. Here, a clear variation of the amount of failures during the day can be distinguished. For example, at 15:00 h, the amount of failures is twice as high as the average (dashed line). At other moments, this amount is only a quarter of the average. This can have large consequences for reliability analysis. As the system load is the largest during the daytime, the impact of component failures during daytime is larger as well. The amount of failures, the hour of the day, and the system load are thus correlated. Normally in reliability analysis, the resulting power system reliability mainly depends on component failures that occur during high system load. Therefore, it would not be surprising if this daily failure pattern leads to a twice as large calculated unreliability of the power system. A possibility to include this correlation into reliability analysis is to vary the component failure frequencies according to this daily pattern.

A separate graph can be made for failures caused by weather effects. The result is shown in the lower graph of Fig. 2.10. It can be seen that failures caused by weather effects are more likely to occur in the evening. The amount of failures in the evening is three times as large as the average. For other failure causes, the NESTOR data does not show significant difference with Fig. 2.10b.

There are various theories to explain the daily pattern of component failures. For example, component failures caused by excavation and hoisting works are likely to occur during working hours. Similarly, failures caused by external causes (third-party damage) are likely to occur during the daytime. Failures caused by weather effects seem to be caused by severe weather effects that mainly occur in the evening in the Netherlands. A causal relationship between failures and the system load can be found in aging and wear, as components are more likely to fail during high loading.

2.3 Historical Blackouts

In the previous section, it was discussed how components of the power system can fail. The failure statistics of these components can be used to determine the reliability of the power system. The methods for this reliability analysis will be described in detail in the following chapters. It is also possible to study the reliability of power systems by analyzing historical blackouts. By investigating which causes and influencing factors led to the blackout, insight is obtained about how component failures can lead to a large blackout. This provides valuable information about which factors should be included in probabilistic reliability analysis in order to make the analysis more accurate and realistic. This section discusses some historical blackouts in the Netherlands and large blackouts in the world.

2.3.1 Large Blackouts

Netherlands: Diemen 2015

On March 27, 2015, at 9:37h, the power supply of about a million customers was interrupted in a large part of the Netherlands, namely the region around Diemen substation [24]. As this substation is part of the 380kV transmission network, a large area was affected, as shown in Fig. 2.11. After investigation, it was found that this blackout was caused by a combination of a technical failure and human misinterpretation during maintenance activities. During maintenance, disconnector A of a substation line bay at Diemen380 was tested. This must be done while the other disconnector (B) of this bay is closed, to maintain a high level of security of supply. One phase of disconnector B was, however, not closed properly because of a mechanical defect. The warning of the control system that disconnector B was not fully closed was misinterpreted as a secondary failure of disconnector B. Based on on-site visual inspection, it was concluded that this disconnector was closed successfully. If both disconnectors A and B are closed, both busbars of the substation are connected through this line bay and both busbars are not protected separately by the protection system. This is an unwanted situation which should last as short as possible. It was therefore decided to ignore the warning of the control system and continue the testing of disconnector A. In reality, one phase of disconnector B was indeed not closed correctly and the opening of disconnector A caused arcing in disconnector B. Influenced by the strong wind, this arcing led to a short circuit with another phase and the protection system switched off the complete substation.

This example illustrates how a combination of technical and human failures can lead to a blackout. Typical in this example is the trade-off between risks. Disconnector A is tested while disconnector B is closed, such that both circuits of the transmission line (of which one circuit is connected to the bay in question) remain in operation for a higher security of supply. However, if both disconnectors are closed, the protection system cannot switch off the busbars of the substation individually in case of a fault. Secondly, the control system indicated that disconnector B was not closed successfully, but it was decided to continue the testing of disconnector A after visual inspection. In this case, a fast decision was needed, as the undesired situation in which both disconnectors are closed should not last too long.

Three factors played a role in this blackout: maintenance activities, a component failure, and human misinterpretation. This example shows that component failures during maintenance can be critical. For reliability analysis, it is therefore recommended to include maintenance in the analysis. Also, it is important to include the behavior of the protection system, as in this specific case the protection system could not switch off the busbars selectively in case of a fault. Furthermore, in this example human failure played a role. Experience shows that human failure played a role in many large blackouts. It would be interesting to include this human factor in (probabilistic) reliability analysis, but it is realized that this remains challenging.

Fig. 2.11 Area affected by the Diemen blackout (map based on [25])

Europe: 2006

On November 4, 2006, a large blackout occurred in Europe [26]. This blackout affected more than 15 million customers and lasted for about two hours. The initial cause of this blackout was a planned disconnection of the transmission line at the Ems river crossing in northwest Germany. This was requested by a shipyard as a large ship needed to pass underneath the transmission line. The neighboring TSOs were informed, and simulations were performed to ensure stability. The planning of the disconnection was, however, changed such that it started two hours earlier. Neighboring TSOs were not informed until late. Full analysis could not be performed, and the sold transfer capacity could not be changed anymore. When the transmission line was disconnected, this caused overloading of other circuits. An attempt to relieve the overload by closing a bus tie worsened the situation instead and caused a cascading outage spreading throughout Europe. Figure 2.12 shows that the European transmission network finally split into three regions because of instability issues. In the region with more production than consumption, over-frequency occurred, which was solved by disconnecting generators and reducing the power generation. In the regions with more consumption than production, under-frequency occurred, which was solved by disconnecting load. After restoring the frequency in all regions, the European transmission network was reconnected in a few attempts.

This blackout illustrates the large impact a cascading outage can have and the events and factors that played a role. Human failures played a prominent role in this

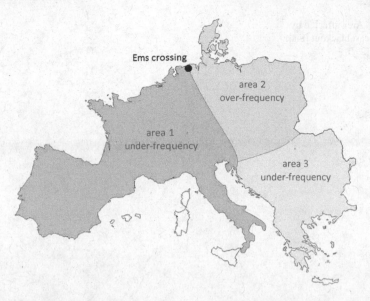

Fig. 2.12 Schematic map of UCTE area split into three areas (map based on [27])

large blackout: Neighboring TSOs were not informed in time about the change in schedule, and TSOs did not have the correct transfer capacities for their grid calculations. Furthermore, the analysis showed that closing the bus tie would relieve the overload, but it was not studied whether this would result in an $N - 1$ situation. As closing the bus tie worsened the overload instead and the situation was not $N - 1$ redundant, this initiated a cascading outage of overloaded connections. It is challenging to reflect this blackout in (probabilistic) reliability analysis. It is relatively easy to include the effect of cascading outages caused by overloaded connections in the analysis, but the main difficulty is the human factor in this blackout. Without the human failures, this blackout would probably not have happened.

USA: Northeast Blackout 2003
On Thursday August 14, 2003, a large blackout occurred in the Northeastern and Midwestern USA and the Canadian province of Ontario [28]. Some power was restored within 7 h, but for many customers, the blackout lasted for two days. For some remote areas, it took a week to get reconnected. The number of affected customers is estimated at 10 million in Ontario and 45 million in eight US states. The blackout was initiated by a power plant that shut down. This led to an overloaded transmission line, but the system operator was not alarmed because of a software bug. The overloaded transmission line heated up, sagged into a tree, and was disconnected by the protection system. Other transmission lines became overloaded, and several of them were automatically disconnected because they sagged into trees as well. The result was a large-scale cascading outage which affected large parts of the USA and Canada, as illustrated in Fig. 2.13. The over-current and undervoltage

Fig. 2.13 Area affected by
the Northeast blackout (map
based on [29])

Blackout area

conditions in this cascading outage led to a large number of transmission lines as
well as production units that were switched off by protection systems.

Although the blackout was initiated by the shut down of a power plant and an
overloaded transmission line, the computer bug enabled this fault to develop into a
large blackout. The computer bug can be considered as a hidden failure as the system
operator was not aware of this failure until a fault occurred. Normally, a transmission
line hitting a tree is a manageable failure, but in this case, it became unmanageable
because of computer failure. This example again illustrates how different factors can
lead to a large blackout. An important factor is the computer bug, but also the system
operator was not completely aware of the vulnerability of the network. Another
factor is the high temperature on that day, which led to a higher system loading (by
air conditioners and fans) and the expansion of overhead lines. For reliability analysis,
it can be of interest to include these factors. For example, a correlation between the
temperature and the failure of overhead lines can be modeled. Computer software
reliability is another topic worth to consider.

2.3.2 Blackouts in the Netherlands

In the Netherlands, the reliability of the power system is reported in yearly reports
called 'Betrouwbaarheid E-Netten' [2–10]. In these reports, the amount of customer

Table 2.7 Interruptions per voltage level (based on [8])

Voltage level	2014	Average (2009–2013)	Circuit length (km)
EHV	0	1	2,760
HV	23	42	9,033
MV	1,837	2,113	104,371
LV	15,897	17,227	223,422
Total	17,757	19,384	339,586

interruptions and the interruption time is reported and compared with the previous years. Also, a description is given of the ten largest blackouts of the year. In the appendices of these report, the failure statistics of network components are shown.

Interruptions in the Netherlands

Table 2.7 is based on the reliability report of 2014 and shows the number of interruptions in the Netherlands. As can be seen, failures in the LV and MV network caused most of the interruptions, while only a few interruptions originated from the EHV and HV network. This is only partly caused by the difference in total circuit length of the specific networks. The main reason is that the lower voltage networks are operated radially, which means that there is only one path from the load to the connection point of the higher voltage network. A connection failure will then lead to the isolation of an area. The customers within such an area can be reconnected by switching actions if possible. The higher voltage networks are operated under redundancy. For example, the EHV network in the Netherlands consists of double circuits and ring structures, creating more redundancy in the network. Failures of components within the EHV network are therefore less likely to cause power interruptions. However, if a failure of the EHV network causes an interruption, a much larger region is affected, as the previously described Diemen blackout illustrates.

Also, Fig. 2.14 shows that customer interruptions are mainly caused by the lower voltage networks. It is interesting that the customer interruption frequency of the MV network is larger than that of the LV network. This is probably caused by the fact that in a failure of the MV network, more customers are involved than in a failure of the LV network.

Ten Largest Blackouts in the Netherlands

Table 2.8 gives an overview of the ten largest blackouts in the Netherlands in the period 2007–2008 and 2010–2016. These blackouts are ranked according to their total *Customer-Interruption-Minutes* (CIM: i.e. the product of the number of affected customers and the average interruption duration). When studying this table in more detail, several causes of power system blackouts can be distinguished. For example, in cases 2 and 10, double circuits are affected by external (third-party) damage. As a result, the areas supplied by the particular double circuits became isolated. In other cases, multiple failures within substations caused isolation of the connected regions. In cases 1, 6, and 8, maintenance activities played a role, while in cases 3

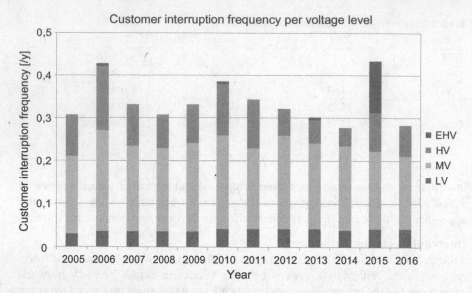

Fig. 2.14 Customer interruption frequency for the Dutch power system (based on [8, 10])

and 7, there was an undesired reaction of the protection system. Case 9 describes a complex combination of failures.

Based on the descriptions of these blackouts, it can again be concluded that large blackouts are often the result of combinations of component failures and influencing factors. To improve (probabilistic) reliability analysis of power systems in the future, these combinations of failures and factors should be included. Examples are dependent failures (common-cause and cascading), maintenance activities and protection system behavior. Some other factors, like human failure and computer software failure, are more difficult to include in the analysis.

2.4   Conclusion

This chapter discussed the failure statistics and failure behavior of power system components and described several historical blackouts. Based on this information, reliability analysis is performed to determine the reliability of power systems. The process of probabilistic reliability analysis is illustratively depicted in Fig. 2.15. On the left, the failure behavior of network components is shown. The figure shows that reliability analysis starts with modeling component failures. These component failures can be dependent (common-cause or cascading) or independent, and are modeled using the failure statistics as described in Sect. 2.1. As discussed in Sect. 2.2, factors like correlation and protection systems influence the power system reliability as well. In order to calculate accurate and realistic results, it might be needed to

Table 2.8 Ten largest blackouts in the Netherlands (2007–2008, 2010–2014) [2–10]

Ranking	C.I.M.	Date	Voltage level (kV)	Location
Description				
1	99,785,104	27-03-2015	380	Diemen
This is the Diemen 2015 blackout as described before in Sect. 2.3.1 in detail				
2	70,749,933	12-12-2007	150	Bommeler-/Tielerwaard
An Apache helicopter hit both circuits of a transmission line; the repair took very long as the lines were broken above a river and the water level was high				
3	16,717,238	9-1-2010	150	Sassenheim
Gas leakage in a GIS caused a short circuit; a rail and a transformer were disconnected; a second transformer was disconnected by protection malfunctioning; and a third one became overloaded				
4	12,687,638	21-6-2016	150	Diemen
Because of a technical defect, one transformer in a substation was disconnected; as this caused fire, the complete installation (incl. second transformer) was switched off for safety				
5	12,335,305	5-1-2013	10	Enschede
While the exact cause was not found after investigation, most probably a short circuit in a MV circuit breaker caused fire and destroyed the MV installation of a substation				
6	9,890,709	30-1-2012	25	Rotterdam
During maintenance works of a secondary installation in a substation, accidentally the wires of a frequency protection were cut and the protection system switched				
7	8,664,054	21-10-2011	150	Tiel & Bommelerwaard
A voltage transformer exploded; because of the protection design and the small fault current, the protection system did not switch selectively; and the complete substation was switched off				
8	8,220,163	14-7-2010	150	Nijmegen
Because of a short circuit (line-earth), a transformer was switched off correctly; a second transformer was under maintenance, such that the third transformer became overloaded				
9	7,935,688	22-6-2012	10	Nieuwegein
This was a complicated outage that occurred after solving a previous outage; a transformer was disconnected after an earth fault; and a second transformer also switched off because of an earth fault				
10	7,630,720	29-3-2007	110	Haaksbergen/Eibergen
A boat hit both circuits of a double-circuit overhead line connection				

include these in probabilistic reliability analysis too. For example, the correlation between the amount of component failures and the time of the day can be reflected by failure frequencies that vary during the day. Furthermore, the failure behavior of protection systems can be included by conditional probabilities that a protection system is not able to clear the fault successfully.

On the right of Fig. 2.15, it is shown that failures eventually can lead to a large blackout. As mentioned in Sect. 2.3, large blackouts are mostly caused by a combination of events where several influencing factors played a role. The cloud in the middle of the figure illustrates the gap between reliability analysis based on

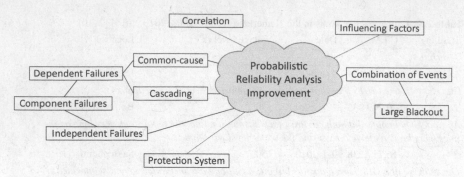

Fig. 2.15 Improvement of probabilistic reliability analysis

component failures and reliability analysis based on the investigation of historical blackouts. Especially here, there is room for improvement of probabilistic reliability analysis. Experience shows that it is hard to calculate the power system reliability as experienced by historical blackouts, as large blackouts are mostly caused by combinations of events, whereas probabilistic reliability analysis usually only considers 'normal' component failures. The accuracy of probabilistic reliability analysis can be improved by including more influencing factors. The analysis of historical blackouts can provide valuable information about which factors should be included. As a first step, the effects of dependent failures, protection system failures, and correlation can be included. Still, it remains challenging to decide which factors must be included, as the causes of large blackouts are diverse.

Problems

2.1 Learning Objectives
Please check the learning objectives as given at the beginning of this chapter.

2.2 Main Concepts
Please check whether you understand the following concepts:

- Failure statistics;
- Failure, failure cause, failure mechanism, failure mode, fault;
- Contingency, contingency analysis;
- (Forced/semi-forced/planned) outage;
- Hidden failure;
- Dependent failure, common-cause, cascading;
- Correlation.

2.3 Failure Statistics
Mention some ways to obtain more accurate failure statistics. What are the disadvantages of these solutions?

2.4 Failure Frequencies EHV/HV
Compare the failure frequencies of EHV and HV components (see Fig. 2.1). What can be said about the failure frequency of EHV and HV overhead lines and transformers? What can be said about the failure frequency of EHV and HV underground cables?

2.5 Repair Times
What are the challenges of defining accurate repair times?

2.6 Failure Progress
Give some examples of failure causes, failure mechanisms, failure modes, and faults.

2.7 Failure Progress Examples
Consider the figure of the progress from failure cause to an outage (Fig. 2.4). Can you mention some real-life examples using this figure?

2.8 Failure Causes
When considering the graphs of the failure causes in this chapter, which measures can you suggest to reduce the number of failures in overhead lines and underground cables?

2.9 Correlation and Dependency
What is the difference between correlation and dependency? If failure statistics show that underground cable failures occur more often during the day, how can we discover dependency/correlation when considering the possible failure causes?

2.10 Failure Modes
What can be said about failure modes of circuit breakers and disconnectors? What are the advantages and disadvantages of defining a failure frequency in /y for these components?

2.11 Protection System
What is the role of the protection system in power system reliability? What are the possible failure modes of a protection system?

2.12 Largest Blackouts Netherlands
Based on the short descriptions in Table 2.8, which cases describe common-cause failures and which cases describe cascading failures? What were the complicating factors?

2.13 Largest Blackouts World
Search on the Internet for a large blackout in the past (e.g. on Wikipedia: List of major power outages). Describe in your own words what series of events led to the blackouts. What was the initial cause and what were the complicating factors?

References

1. TenneT TSO (2011) NESTOR (Nederlandse Storingsregistratie) database 2006–2010. TenneT TSO B.V, Arnhem, Netherlands
2. Nederland Enbin (2008) Betrouwbaarheid van Elektriciteitsnetten in Nederland - Resultaten 2007. Enbin Nederland, The Netherlands
3. Nederland Netbeheer (2009) Betrouwbaarheid van Elektriciteitsnetten in Nederland - Resultaten 2008. Netbeheer Nederland, The Netherlands
4. Nederland Netbeheer (2010) Betrouwbaarheid van Elektriciteitsnetten in Nederland - Resultaten 2010. Netbeheer Nederland, The Netherlands
5. Nederland Netbeheer (2012) Betrouwbaarheid van Elektriciteitsnetten in Nederland - Resultaten 2011. Netbeheer Nederland, The Netherlands
6. Nederland Netbeheer (2013) Betrouwbaarheid van Elektriciteitsnetten in Nederland - Resultaten 2012. Netbeheer Nederland, The Netherlands
7. Nederland Netbeheer (2014) Betrouwbaarheid van Elektriciteitsnetten in Nederland - Resultaten 2013. Netbeheer Nederland, The Netherlands
8. Nederland Netbeheer (2015) Betrouwbaarheid van Elektriciteitsnetten in Nederland - Resultaten 2014. Netbeheer Nederland, The Netherlands
9. Nederland Netbeheer (2016) Betrouwbaarheid van Elektriciteitsnetten in Nederland - Resultaten 2015. Netbeheer Nederland, The Netherlands
10. Nederland Netbeheer (2017) Betrouwbaarheid van Elektriciteitsnetten in Nederland - Resultaten 2015. Netbeheer Nederland, The Netherlands
11. VDN (2004) Ermittlung von Eingangsdaten für Zuverlässigkeitsberechnungen aus der VDN-Störungsstatistik. VDN, Germany. http://www.vde.com/de/fnn/arbeitsgebiete/versorgungsqualitaet/seiten/versorgungszuverlaessigkeit-unterlagen.aspx
12. FNN (2013) Ermittlung von Eingangsdaten für Zuverlässigkeitsberechnungen aus der FNN-Störungsstatistik. FNN, Germany
13. Cigré Working Group B1.10 (2009) Update of service experience of HV underground and submarine cable systems, TB379. Cigré, France
14. Meijer S, Smit J, Chen X, Fischer W, Colla L (2011) Return of experience of 380 kV XLPE landcable failures. Jicable - 8th international conference on insulated power cables, Versailles, France
15. Cigré Working Group A3.06 (2012) Reliability of HV equipment and GIS. http://www.mtec2000.com/cigre_a3_06/.htm
16. Tuinema BW, Rueda JL, van der Sluis L, van der Meijden MAMM (2015) Reliability of transmission links consisting of overhead lines and underground cables. IEEE Trans Power Deliv 31(3):1251–1260
17. Rausand M, Hoyland A (2004) System reliability theory: models and statistical methods. Wiley, Canada
18. Standard British (1979) Glossary of terms used in quality assurance including reliability and maintainability terms. British Standards Institution, London
19. Klaassen KB, van Peppen JCL, Bossche A (1988) Bedrijfszekerheid. Delftse Uitgevers Maatschappij, Delft, the Netherlands, Theorie en Techniek
20. Bossche A (2016) Microelectronics reliability - lecture notes. Delft University of Technology, Delft, the Netherlands
21. International Electrotechnical Commission (1997) Functional safety of electrical/electronic/programmable electronic safety-related systems - Part 1–7. International Electrotechnical Commission, Geneva, Switzerland
22. Grainger JJ, Stevenson WD Jr (1994) Power system analysis. McGraw-Hill International Editions, New York
23. Li W (2005) Risk assessment of power systems - models, methods, and applications. Wiley Interscience - IEEE Press, Canada
24. TenneT TSO B.V. (2015) Samenvatting, conclusies en aanbevelingen storing Diemen. TenneT TSO B.V., Arnhem, The Netherlands

25. d-maps.com (2018) Free map of the Netherlands. http://d-maps.com/carte.php?num_car=15029&lang=en

26. Wikipedia (2016) 2006 European blackout. https://en.wikipedia.org/wiki/2006_European_blackout

27. d-maps.com (2018) Free map of Europe. http://d-maps.com/carte.php?num_car=2233&lang=en

28. Wikipedia (2016) Northeast blackout of 2003. https://en.wikipedia.org/wiki/Northeast_blackout_of_2003

29. d-maps.com (2018) Free map of North America. http://d-maps.com/carte.php?num_car=1405&lang=en

Part II
Modeling

Chapter 3
Reliability Models of Components

Probabilistic reliability analysis starts with accurate modeling of the individual components within a system. In the previous chapter, it was explained how power systems can fail, while the failure process and failure statistics of power system components were discussed in detail as well. This information is used for modeling the reliability behavior of individual components within the power system. Several methods for component modeling are discussed in this chapter, starting with a description of the basic reliability functions in Sect. 3.1. Then, in Sect. 3.2, the life cycle of repairable components is discussed. The two-state Markov model is also often used to model the reliability of individual components and is described in Sect. 3.3, while the stress-strength model is shortly introduced in Sect. 3.4. General conclusions are summarized in Sect. 3.5.

Learning Objectives

After reading this chapter, the student should be able to:

- describe how the reliability of power system components is modeled;
- use and derive basic reliability functions;
- explain the meaning of the bathtub curve and how it is used;
- understand the component life cycle and derive related equations;
- model the behavior of a component with a (two-state) Markov model.

3.1 Reliability Functions

In reliability analysis of components, a distinction can be made between *repairable components* and *unrepairable components*. The concept unrepairable component should be understood in a broad sense, as some components are considered 'unrepairable' because of economic reasons only. Unrepairable components can therefore,

© Springer Nature Switzerland AG 2020

B. W. Tuinema et al., *Probabilistic Reliability Analysis of Power Systems*,
https://doi.org/10.1007/978-3-030-43498-4_3

for example, be the components of a satellite, but also the computer chips within a laptop. Repairable components are repaired after a failure. Often, these components are regularly maintained as well. If repair of the failed component is not possible or economically unattractive, the component is replaced by a new one, unless it is not needed anymore.

3.1.1 Basic Reliability Functions

To describe the failure behavior of components without repair, several basic reliability functions are defined [1, 2]:

- $F(t)$: *unreliability function* or *failure distribution*
  The probability of finding an originally healthy component in a failed state after time t. Assuming that t_{failure} is the moment of failure, the unreliability is described as:

$$F(t) = P[t_{\text{failure}} \leq t] \tag{3.1}$$

The unreliability thereby is the probability that the component has failed before (or at) time t. Normally, it is assumed that the component was working at $t = 0$ and the component finally has failed at $t = \infty$, resulting in: $F(0) = 0$ and $F(\infty) = 1$. The unreliability function is a continuously increasing function. The unreliability function is in fact a cumulative distribution function (CDF) of the component failures (see also Appendix A).

- $R(t)$: *reliability function*
  The probability of finding an originally healthy component in a healthy state after time t. The reliability function can be calculated directly from the unreliability function, as the sum of both must be 1 (i.e. either the component has failed before time t or it is still healthy and will fail after time t). Consequently:

$$R(t) = P[t_{\text{failure}} > t] = 1 - P[t_{\text{failure}} \leq t] = 1 - F(t) \tag{3.2}$$

The reliability function can be seen as 1 minus the CDF of the component failures.

- $f(t)$: *failure density distribution*
  The rate at which a component fails at time t. The failure density distribution is the derivative of the failure distribution. If $F(t)$ can be differentiated:

$$f(t) = \lim_{\Delta t \to 0} \frac{1}{\Delta t} P[t < t_{\text{failure}} \leq t + \Delta t] = \lim_{\Delta t \to 0} \frac{F(t + \Delta t) - F(t)}{\Delta t} = \frac{dF(t)}{dt} \tag{3.3}$$

The failure density distribution is therefore the probability density function (PDF) of the component failures (see also Appendix A).

Table 3.1 Relation between reliability functions (based on [2])

Function of:	$f(t)$	$F(t)$	$R(t)$	$h(t)$
$f(t) =$	$f(t)$	$\frac{\mathrm{d}F(t)}{\mathrm{d}t}$	$-\frac{\mathrm{d}R(t)}{\mathrm{d}t}$	$h(t)\mathrm{e}^{-\int_0^t h(t)\mathrm{d}t}$
$F(t) =$	$\int_0^t f(t)\mathrm{d}t$	$F(t)$	$1 - R(t)$	$1 - \mathrm{e}^{-\int_0^t h(t)\mathrm{d}t}$
$R(t) =$	$\int_t^\infty f(t)\mathrm{d}t$	$1 - F(t)$	$R(t)$	$\mathrm{e}^{-\int_0^t h(t)\mathrm{d}t}$
$h(t) =$	$\frac{f(t)}{\int_t^\infty f(t)\mathrm{d}t}$	$\frac{\frac{\mathrm{d}F(t)}{\mathrm{d}t}}{1-F(t)}$	$-\frac{\mathrm{d}(\ln R(t))}{\mathrm{d}t}$	$h(t)$

- $h(t)$: *hazard rate*
  The rate at which a component fails at time t, given that it is still healthy at time t. The hazard rate can be calculated as follows:

$$
\begin{aligned}
h(t) &= \lim_{\Delta t \to 0} \frac{1}{\Delta t} P[t < t_{\text{failure}} \le t + \Delta t \,|\, t_{\text{failure}} > t] \\
&= \lim_{\Delta t \to 0} \frac{1}{\Delta t} \frac{P[t < t_{\text{failure}} \le t + \Delta t]}{P[t_{\text{failure}} > t]} \\
&= \lim_{\Delta t \to 0} \frac{F(t + \Delta t) - F(t)}{\Delta t\, R(t)} = \frac{\mathrm{d}F(t)}{\mathrm{d}t} \frac{1}{R(t)} = \frac{f(t)}{R(t)}
\end{aligned}
\tag{3.4}
$$

The hazard rate is actually a conditional PDF of the component failures.

An example of applying the reliability functions is given in Example 3.1. Table 3.1 shows the relation between the different reliability functions.

From the reliability functions, the expected lifetime of the component can be calculated. This is the expected value of the failure density distribution [2]:

$$
\theta = \int_0^\infty t f(t)\mathrm{d}t
\tag{3.5}
$$

The expected lifetime is also the area below the reliability function:

$$
\theta = \int_0^\infty R(t)\mathrm{d}t
\tag{3.6}
$$

This can be derived as follows:

$$
\begin{aligned}
\theta &= \int_0^\infty R(t)\mathrm{d}t = \int_0^\infty \int_t^\infty f(x)\mathrm{d}x\mathrm{d}t = \int_0^\infty \int_0^x f(x)\mathrm{d}t\mathrm{d}x \\
&= \int_0^\infty [t f(x)]_0^x \,\mathrm{d}x = \int_0^\infty x f(x)\mathrm{d}x
\end{aligned}
\tag{3.7}
$$

which is essentially the same as (3.5).

Example 3.1: Reliability Functions

Question Assume that we have 10 components of a special type. The first four components fail after 1, 2, 3 and 4 years respectively. The failed components are not repaired. The next 3 components fail with one every other year, i.e. after 6, 8 and 10 years respectively. The remaining 3 components then fail with one every three years. Thus, the components fail after 1, 2, 3, 4, 6, 8, 10, 13, 16, 19 years respectively. Based on this information, draw the reliability functions of this specific component type. Also, calculate the expected lifetime of this component type.

Answer First, we start with the reliability function ($R(t)$). At time $t = 0$, all components are still working, so the reliability is $R(0) = 1$. Then, after one year, one component fails. As 90% of the components are still working then, the reliability function becomes $R(1) = 0.9$. After two years, a second component fails. As 80% of the components are still working then, the reliability function becomes $R(2) = 0.8$. This continues for the other components, until all components have failed after 19 years ($R(19) = 0$). Table 3.2 shows the values of the reliability function. The reliability function is also shown in Fig. 3.1.

The next step is the calculation of the unreliability function. As the unreliability function is $F(t) = 1 - R(t)$, it can be calculated directly from the reliability function. The values are shown in Table 3.2 and Fig. 3.1 as well.

The failure density distribution function is the derivative of the unreliability function. Therefore, we can calculate that it is:

$$f(t) = \frac{0.4 - 0}{4 - 0} = \frac{1}{10} /y \qquad \text{for } 0 \leq t < 4;$$

$$f(t) = \frac{0.7 - 0.4}{10 - 4} = \frac{1}{20} /y \qquad \text{for } 4 \leq t < 10;$$

$$f(t) = \frac{1 - 0.7}{19 - 10} = \frac{1}{30} /y \qquad \text{for } 10 \leq t < 19;$$

$$f(t) = 0 /y \qquad \text{for } 19 \leq t. \qquad (3.8)$$

Note that this function is not continuous. The values of the failure density distribution function are shown in Table 3.2, while a sketch of the failure density distribution function can be found in Fig. 3.1.

The hazard rate can be calculated from the failure density function by pointwise dividing the values, as $h(t) = f(t)/R(t)$. The values of the hazard rate then become as shown in Table 3.2. The graph of the hazard rate can be found in Fig. 3.1. Note that this function is not continuous as well and it increases when time increases. Because there are less components left when time increases, the remaining components must fail faster to observe a constant number of

Table 3.2 Reliability functions of the considered component (Example 3.1)

t [y]	$R(t)$ [-]	$F(t)$ [-]	$f(t)$ [/y]	$h(t)$ [/y]
0	1	0	1/10	0.10
1	0.9	0.1	1/10	0.11
2	0.8	0.2	1/10	0.13
3	0.7	0.3	1/10	0.14
4	0.6	0.4	1/20	0.08
6	0.5	0.5	1/20	0.10
8	0.4	0.6	1/20	0.13
10	0.3	0.7	1/30	0.11
13	0.2	0.8	1/30	0.17
16	0.1	0.9	1/30	0.33
19	0	1	0	–

failures per time for the complete group of components. At $t \rightarrow 19$, the hazard rate increases to infinity and at $t \geq 19$, the hazard rate is undefined as it is $0/0$.

The expected lifetime of this component type equals the area below the reliability function. By summing the areas of the triangles and squares below the reliability function:

$$
\begin{aligned}
\theta &= \int_0^\infty R(t)\mathrm{d}t \\
&= \frac{1}{2} \cdot 4 \cdot (1 - 0.6) + 4 \cdot 0.6 + \frac{1}{2} \cdot (10 - 4) \cdot (0.6 - 0.3) \\
&\quad + (10 - 4) \cdot 0.3 + \frac{1}{2} \cdot (19 - 10) \cdot 0.3 \\
&= \frac{1}{2} \cdot 4 \cdot 0.4 + 4 \cdot 0.6 + \frac{1}{2} \cdot 6 \cdot 0.3 + 6 \cdot 0.3 + \frac{1}{2} \cdot 9 \cdot 0.3 \\
&= 0.8 + 2.4 + 0.9 + 1.8 + 1.35 = 7.25\,\text{y}
\end{aligned}
\tag{3.9}
$$

The approach described in Example 3.1 is actually the main approach to find the reliability functions of specific component types. Failure statistics can be obtained by testing of components or can be extracted from databases with historical failures of components in operation. Then, based on these statistics, a specific probability distribution is fit to the data.

For example, in a new office building with 1000 light bulbs with an expected lifetime of 2 years, the occurrence of light bulb failures can be approximated by the *normal distribution*. This because of the central-limit theorem [3]. The failure density distribution then is [4]:

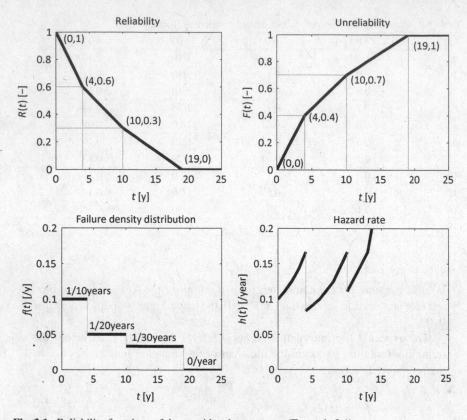

Fig. 3.1 Reliability functions of the considered component (Example 3.1)

$$f(t) = \frac{1}{\sigma\sqrt{2\pi}}e^{-\frac{(t-\mu)^2}{2\sigma^2}} \qquad (3.10)$$

where
t = time
μ = mean
σ = standard deviation

Figure 3.2 shows the number of light bulb failures in the office building as approximated by the normal distribution. The mean and standard deviation can be determined based on test results or experience with failures of these specific light bulbs in reality. In this example, it is assumed that $\mu = 24$ months and $\sigma = 2$ months.

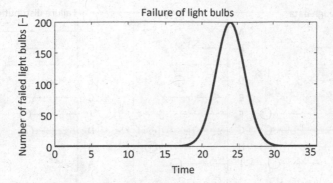

Fig. 3.2 Failure of light bulbs in an office building

Example 3.2: Data Analysis

In Example 3.1, it was discussed how the reliability functions can be derived from component failure data. In the ideal case, all required failure information (like the year of installation, the time of failure and the operational time) would be available. Unfortunately, this is often not the case in reality. A distinction can be made here between *uncensored data* and *censored data*. For uncensored data, it is exactly known when the component was installed and when it failed. For censored data, this is not known. For *left censored data*, the installation time is not known and for *right censored data*, the time of failure is unknown.

Figure 3.3 illustrates the process of data analysis [5]. As can be seen, component failures are observed during a certain observation period. In the ideal case, we know exactly when the component was installed and when it failed. This is the case for components A and B.

Some components were installed before the observation period started. This is the case for components C and D. For component C, it is known that the installation was exactly Δt before the observation period started. For this component, the failure data must be normalized by adding Δt to its lifetime during the observation period. For component D, it is not known how long it had been in operation before the observation period started. For this component, we only can say that it survived longer than its lifetime during the observation period. In the middle figure, this is indicated with a circle at the end of its lifetime. Component E was installed after the observation period started. The lifetime of this component must therefore be normalized by shifting it to the left.

Some components survived the observation period. This is the case for components F, G and H. Component F was installed when the observation period started. Component G was installed before the observation period started, but it is unknown when it was installed. It can therefore only be concluded that it

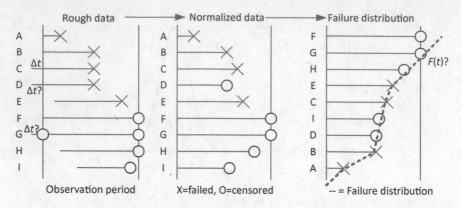

Fig. 3.3 Data analysis of the failure data (Example 3.2)

survived the observation period, such that the normalized lifetime is similar to that of component F. Component H was installed after the observation period started. The lifetime of this component is therefore shifted to the left. Component I was installed after the observation period started, but was taken out of operation without failing before the observation period ended. The lifetime of this component is shifted to the left.

The normalized data can now be ordered to estimate the failure distribution [6]. Here, the longest lifetimes are placed on top and the shorter lifetimes below. The failure distribution is found in a way comparable to Example 3.1. The censored data however complicates this process. As the exact lifetime of these components is unknown, the exact order of the components is unknown. A solution is to consider every possible component order and estimate the failure distribution [7]. Based on the results, the best approximation of the failure distribution is determined by a computer program. Confidence bounds can be given to indicate the accuracy of the approximation.

3.1.2 Bathtub Curve

When studying the failure behavior of components in practice, it is observed that the hazard rate often follows the so-called *bathtub curve* as illustrated in Fig. 3.4. According to the bathtub curve, a component experiences an *infant stage* with a decreasing hazard rate first. In this infant stage, the hazard rate is larger because some components are still vulnerable to early failures like production failures. These kinds of failures are likely to occur in the first stage of the component lifetime.

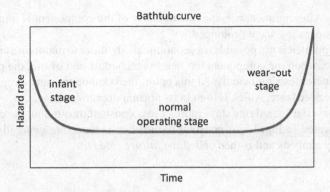

Fig. 3.4 Bathtub curve of component failures

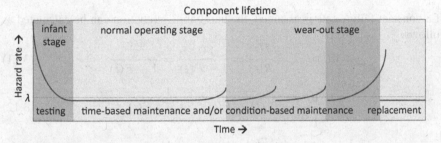

Fig. 3.5 Bathtub curve with asset management and definition of failure rate

After the infant stage, a *normal operating stage* follows, in which the hazard rate stays constant. In this stage, failures occur more or less randomly. Finally, the component reaches the *wear-out stage*, in which the hazard rate increases significantly. In this stage, the component reaches its end of life and failures like wear-out failures are very likely to occur.

In asset management of power systems, the bathtub curve is an important concept as it can be influenced in various ways to optimize the reliability of components. This is illustrated in Fig. 3.5. First, the higher hazard rate in the infant stage must be avoided. In practice, this is done by testing the components. By performing tests on the components, the first stage of component operation is simulated such that the components are in the normal operating stage when they are put into service. To save time, *accelerated tests* can be performed, e.g. by stressing the components under a higher voltage than normal.

During the normal operating stage, the functionality of the component is assured by performing regular *maintenance*. Maintenance is any human intervention to keep a component (or system) in a healthy state or bring it back to a healthy state [1]. Maintenance can be *time-based maintenance*, where the maintenance is performed at regular time instants, or it can be *condition-based maintenance*, where maintenance is performed when the condition of the component deteriorates and failures become more likely. After maintenance, the functionality of the component is improved and the normal operating stage prolonged.

At some point, it is not possible (or economical) anymore to maintain a component. For example, when the component has nearly reached its end of life, the probability of failures increases significantly. At this point, the component can be replaced with a new (and tested) one, which is then in its normal operating stage.

In this way, the hazard rate stays more or less constant throughout the component life, as indicated in Fig. 3.5. A constant hazard rate is therefore generally assumed in reliability analysis and is then called the *failure rate* (λ).

3.1.3   Negative Exponential Distribution

If a constant failure rate is assumed, the unreliability function can be calculated as follows:

$$h(t) = \lambda = \frac{f(t)}{R(t)} = \frac{f(t)}{1 - F(t)} = \frac{\frac{dF(t)}{dt}}{1 - F(t)} \tag{3.11}$$

$$\frac{dF(t)}{dt} = \lambda(1 - F(t)) = \lambda - \lambda F(t) \tag{3.12}$$

$$\frac{dF(t)}{dt} + \lambda F(t) = \lambda \tag{3.13}$$

The solution of this last differential equation is:

$$F(t) = 1 - e^{-\lambda t} \tag{3.14}$$

The reliability functions then become [1]:

$$h(t) = \lambda \tag{3.15}$$

$$F(t) = 1 - e^{-\lambda t} \tag{3.16}$$

$$R(t) = 1 - F(t) = e^{-\lambda t} \tag{3.17}$$

$$f(t) = \frac{dF(t)}{dt} = \lambda e^{-\lambda t} \tag{3.18}$$

In Fig. 3.6, the graphs of these reliability functions are shown. The probability distribution of the reliability function is known as the *(negative) exponential distribution*. As generally a constant hazard rate (i.e. the failure rate) is assumed, this exponential distribution is often used in reliability analysis. The exponential distribution has some characteristics that make further reliability analysis easier.

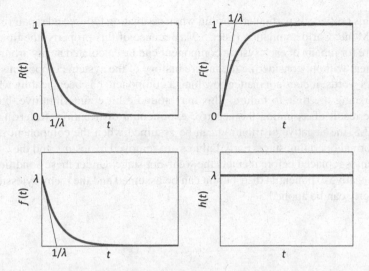

Fig. 3.6 Reliability functions according to the (negative) exponential distribution

For the exponential distribution, the expected component lifetime can be calculated as follows (using the integration rules of exponential functions):

$$\theta = \int_0^\infty tf(t)dt = \int_0^\infty t\lambda e^{-\lambda t}dt = \lambda \left[\frac{te^{-\lambda t}}{-\lambda} - \frac{1}{-\lambda} \int e^{-\lambda t}dt \right]_0^\infty$$

$$= \lambda \left[-\frac{te^{-\lambda t}}{\lambda} - \frac{1}{\lambda^2}e^{-\lambda t} \right]_0^\infty = \frac{1}{\lambda} \tag{3.19}$$

This means that if a specific component has a failure rate of $\lambda = \frac{1}{10}/y$, its expected lifetime is $\theta = 10\,y$. Another characteristic of the exponential distribution is that the tangential of the reliability function at $t = 0$ crosses the x-axis at $1/\lambda$, exactly the expected average lifetime of the component. In this way, the expected lifetime can be read directly from the graphs of the reliability functions. This is also illustrated in Fig. 3.6.

Example 3.3: Memorylessness

A typical characteristic of the exponential distribution is that it is *memoryless*, i.e. for a given time t_0, the failure distribution of $t + t_0$ given that the component is healthy at t_0 is the same as the failure distribution of t. This can be proved as follows:

$$f(t + t_0|\text{healthy at } t_0) = \frac{f(t + t_0)}{R(t_0)} = \frac{\lambda e^{-\lambda(t+t_0)}}{e^{-\lambda t_0}} = \lambda e^{-\lambda t} = f(t) \tag{3.20}$$

This property is particularly useful when estimating failure and repair times in a Monte Carlo simulation (Sect. 5.3). Because of this property, the time to failure (or repair) of each system component can be calculated at any arbitrary moment without considering the failure history of the system components. In others words, it does not matter how long a component has been healthy when calculating the time to failure. This may sound a bit counterintuitive, but it is the result of assuming the negative exponential distribution. Referring to Fig. 3.5, the negative distribution can be assumed when the component is in its normal operating stage. Early failures are removed by testing and the component is replaced before it enters the wear-out stage. Under these conditions, the negative exponential distribution can be assumed and the memorylessness property can be applied.

3.1.4 Weibull Distribution

Previously, it was described how a constant hazard rate during the normal operating stage of a component is modeled. In reliability analysis, especially in asset management, it can be of interest to model the decreasing hazard rate during the infant stage and the increasing hazard rate during the wear-out stage. One probability distribution is particularly useful for this purpose: the *Weibull distribution*. The Weibull distribution is a transformation from the exponential distribution and is often used for curve fitting of probability distributions.

According to the Weibull distribution, the reliability functions are defined as [4]:

$$R(t) = e^{-\left(\frac{t}{\alpha}\right)^c} \tag{3.21}$$

$$F(t) = 1 - e^{-\left(\frac{t}{\alpha}\right)^c} \tag{3.22}$$

$$f(t) = \frac{c}{\alpha} \left(\frac{t}{\alpha}\right)^{c-1} e^{-\left(\frac{t}{\alpha}\right)^c} \tag{3.23}$$

$$h(t) = \frac{c}{\alpha} \left(\frac{t}{\alpha}\right)^{c-1} \tag{3.24}$$

where
c = shape parameter ($c \geq 0$)
α = scale parameter ($\alpha \geq 0$)

When comparing these equations to the reliability functions of the exponential distribution (Sect. 3.1.3), some similarities can be found. For example, when the shape parameter $c = 1$, the Weibull distribution equals the exponential distribution, with $\frac{1}{\alpha} = \lambda$. For a shape parameter $c < 1$, the hazard rate decreases over time. And for a shape parameter $c > 1$, the hazard rate increases with time.

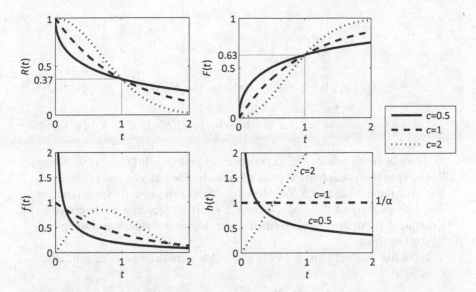

Fig. 3.7 Reliability functions according to the Weibull distribution

In Fig. 3.7, the reliability functions according to the Weibull distribution are shown. The graphs show that the hazard rate indeed decreases with time if $c < 1$. This is particularly useful for modeling the hazard rate during the infant stage of a component. For $c > 1$, the hazard rate increases, which can be used to model the wear-out stage of components.

Other distributions used for reliability analysis of components are the *lognormal distribution*, the *gamma distribution*, the *Poisson distribution* and the *Rayleigh distribution* [1, 4].

Example 3.4: Weibull Analysis

The Weibull distribution is often used for failure data analysis of components. It is then first assumed that the failure distribution of a particular component follows the Weibull distribution. As a second step, the shape and scale parameters of the Weibull distribution are estimated.

For this Weibull analysis, it is helpful if the Weibull distribution can be drawn as a straight line. Based on (3.21) [7, 8]:

$$F(t) = 1 - e^{\left(\frac{t}{\alpha}\right)^c} \tag{3.25}$$

$$\left(\frac{t}{\alpha}\right)^c = -\ln(1 - F(t)) \tag{3.26}$$

$$\ln\left(\left(\frac{t}{\alpha}\right)^c\right) = \ln\left(-\ln(1 - F(t))\right) \tag{3.27}$$

$$c \ln t = c \ln \alpha + \ln(-\ln(1 - F(t))) \tag{3.28}$$

$$\ln t = \ln \alpha + \frac{1}{c}\ln(-\ln(1 - F(t))) \tag{3.29}$$

There is now a linear relation between $\ln t$ and $\ln(-\ln(1 - F(t)))$. Using Weibull paper, in which $\ln t$ is plotted on the horizontal axis and $\ln(-\ln(1 - F(t)))$ on the vertical axis, a Weibull distribution can now be drawn as a straight line. The plotting of the graph follows a similar procedure as described in Example 3.1, but now using Weibull paper. Figure 3.8 shows an example of Weibull analysis.

From the Weibull plot, the scale and shape parameters can be estimated. For $t = \alpha$:

$$F(t = \alpha) = 1 - e^{\left(\frac{\alpha}{\alpha}\right)^c} = 1 - \frac{1}{e} = 0.63 \tag{3.30}$$

So, at $F(t) = 0.3$, $t = \alpha$. From the graph in Fig. 3.8, it can be read that $\alpha = 4$. The shape parameter can now be calculated:

$$c = \frac{\ln(-\ln(1 - F(t)))}{\ln t - \ln \alpha} \tag{3.31}$$

For example, using the point $(1, 0.07)$ in the graph:

$$c = \frac{\ln(-\ln(1 - 0.07))}{\ln 1 - \ln 4} = 1.9 \tag{3.32}$$

As the shape parameter $c > 1$, it can be concluded that the hazard rate of this particular component increases with time.

In a study on the failures of EHV cable terminations, a Weibull analysis was performed [9, 10]. As shown in Fig. 3.9, a decreasing hazard rate was found for these components. Based on this result and the fact that EHV underground cables are a relatively new technology, it was concluded that EHV terminations are still in the infant stage of their lifetime. It was therefore concluded that the failure frequency of these terminations will decrease in the coming years.

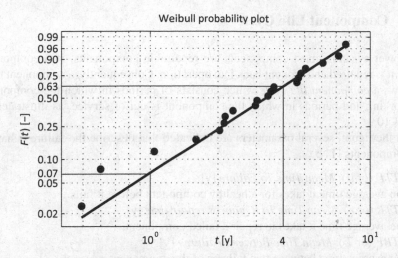

Fig. 3.8 Example of a Weibull analysis (Example 3.4)

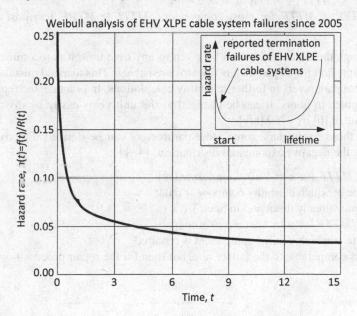

Fig. 3.9 Weibull analysis of termination failures (Example 3.4) (based on [9])

3.2 Component Life Cycle

In power systems, most components are repairable components. If a component fails, it is repaired and brought back into operation. Consequently, a component then follows a *component life cycle* which consists of periods in which the component is working and periods in which the component is out of service, as illustrated in Fig. 3.10.

In the figure, several parameters are indicated that describe the failure behavior of components [11]:

- *MTTF* (or *d*): *Mean Time To Failure* [y]
  The average time it takes for a healthy component before it fails.
- *MTTR* (or *r*): *Mean Time To Repair* or *repair time* [y]
  The average time it takes to repair a failed component.
- *MTBF* (or *T*): *Mean Time Between Failures* [y]
  The average time between two failures of the component.
  ($MTBF = MTTF + MTTR$ or $T = d + r$)
  (as $MTTR \ll MTTF$ for most components, $MTTF \approx MTBF$ for most components.)

Although these parameters can be given in any time unit, it is recommended to use the same unit for all parameters to keep consistency. This to avoid mistakes when using these parameters in further reliability calculations. In practice, the repair time is often given in hours. It can be changed to the unit years easily by dividing the repair time in [h] by 8760 [h/y].

From these parameters, some other parameters can be derived directly (when assuming the negative exponential distribution) [4, 11]:

- $\lambda = 1/MTTF$ (or $\lambda = 1/d$): *failure rate* [/y]
  The rate at which a healthy component fails.
  (This was already discussed in Sect. 3.1.3.)
- $\mu = 1/MTTR$ (or $\mu = 1/r$: *repair rate* [/y]
  The rate at which a failed component is repaired.
  (This is comparable to the failure rate, but then for the repair process.)

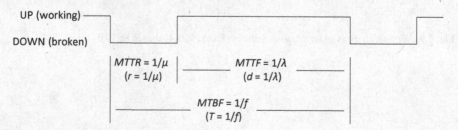

Fig. 3.10 Life cycle of a component

- $f = 1/MTBF$ (or $f = 1/T$: *failure frequency* [/y]
  The average frequency at which a component fails.
  (As $MTTR \ll MTTF$ (or $r \ll d$) for most components, $f \approx \lambda$ for most components.)

Example 3.5: The Repair Process

Previously, it was discussed how the failure process of components can be modeled by the exponential distribution. The same exponential distribution can be used to model the repair process of components. The counterpart of the failure rate λ is then called the repair rate μ, which describes the rate at which the component is repaired.

The exponential distribution was used to model the repair time of overhead lines in the Dutch power system, as illustrated in Fig. 3.11. On the x-axis, the repair time is shown, while the y-axis shows the amount of components that have not been repaired yet. At repair time 0, all failed components are still unrepaired and at repair time 50, most of the components have been repaired.

First, with Matlab's curve fitting tool [12], an exponential distribution was fit to the repair data. As the graph shows, this exponential distribution does not approximate the repair process very well. Clearly, it is not possible to model both the shorter repairs as well as the longer repairs with a single exponential distribution. A study on German failure statistics showed that the repair process of power system components can better be approximated by a combination of two exponential distributions [13]:

$$c_1 e^{-\frac{1}{T_1}} + c_2 e^{-\frac{1}{T_2}} \tag{3.33}$$

This was also done for the repair data of overhead lines from the Dutch NESTOR database [14]. As Fig. 3.11 shows, the approximation by a combination of two exponential distributions represents the repair process very well. If the graph is studied in more detail, still some extremely long repairs are not covered by the combination of two exponential distributions. Extremely long repair times can occur, for example, when a tower has collapsed and needs to be rebuilt. These extremely long repair times are sometimes modeled separately in reliability analysis.

Although the repair process can be modeled more accurately by a combination of multiple exponential distributions, a single exponential approximation is generally preferred for further reliability analysis. The main reason is that modeling the failure behavior of individual components by combinations of multiple exponentials easily results in too complicated models of large systems. For simplification purposes, the single exponential distribution is thus preferred, even though it is an inferior approximation.

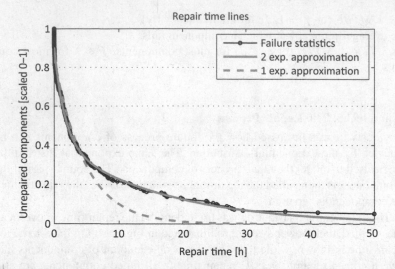

Fig. 3.11 Exponential approximation of repair times (Example 3.5)

In Sect. 2.1, the repair times of several power system components as derived from historical failure statistics were discussed. Table 2.3 gave an overview of the repair times. It was mentioned that one must take care that the repair times are modeled accurately. After studying this example, the difference between average, 3rd-quartile and exponential approximation are more clear now. For example, the statistical mean value of the repair times in this example was 117 h. However, from the boxplot as shown in Fig. 3.12, it can be seen that 75% of the overhead line failures were repaired within 8 h. The approximation with a single exponential showed an average repair time of 4 h. Simply using the statistical mean value in further reliability analysis would probably lead to inaccurate and too pessimistic results, while the 3rd-quartile value is more comparable to the repair time found by the single exponential approximation.

With respect to the component life cycle, two more concepts are introduced [11]:

- $A = d/T$: *availability* [-]
  The probability that a component is found in a healthy state at an arbitrary time
  (or: The average fraction of time that a component is in a healthy state.)
- $U = r/T$: *unavailability* [-]
  The probability that a component is found under repair at an arbitrary time
  (or: The average fraction of time that a component is under repair.)

Fig. 3.12 Boxplot of the repair times of overhead lines (Example 3.5)

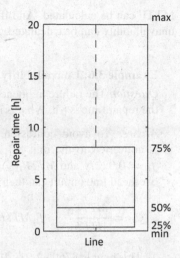

The availability and unavailability can be calculated as follows [11, 15]:

$$A = \frac{MTTF}{MTBF} = \frac{MTTF}{MTTF + MTTR} \tag{3.34}$$

$$U = \frac{MTTR}{MTBF} = \frac{MTTR}{MTTF + MTTR} \tag{3.35}$$

From these equations, other equations that relate the previously described parameters can be derived:

$$A = \frac{MTTF}{MTBF} = \frac{f}{\lambda} \tag{3.36}$$

$$A = \frac{MTTF}{MTTF + MTTR} = \frac{\frac{1}{\lambda}}{\frac{1}{\lambda} + \frac{1}{\mu}} = \frac{\mu}{\lambda + \mu} \tag{3.37}$$

$$U = \frac{MTTR}{MTBF} = f \cdot MTTR \tag{3.38}$$

$$U = \frac{MTTR}{MTTF + MTTR} = \frac{\frac{1}{\mu}}{\frac{1}{\lambda} + \frac{1}{\mu}} = \frac{\lambda}{\lambda + \mu} \tag{3.39}$$

When using these equations, one must take care that the right units are used. In all these equations, it is assumed that MTTF, MTTR, and MTBF are measured in [y], and λ, μ and f in [/y]. If one of the parameters is measured in another unit (e.g. MTTR is often given in [h]), this must be converted to the right unit (e.g. by dividing the MTTR in [h] by 8760 [h/y]). Moreover, only two parameters can be independent in each equation. For example, if the failure frequency and repair time are given, the

MTTF can be calculated. And if the availability and the failure rate are given, the unavailability can be calculated.

Example 3.6: Unavailability Overhead Line

Question The failure frequency of a 10 km overhead line circuit is 0.022 /y and the repair time is 8 h. What is its unavailability? How many hours/year is this?

Answer To avoid confusion, the repair time of the overhead line circuit is converted to the unit [y], making the failure frequency and repair time $f_{11} = 0.022/\text{y}$ and $MTTR_{11} = \frac{8}{8760} = 9.13 \cdot 10^{-4}$ y. The unavailability of the overhead line circuit can then be calculated as:

$$U_{11} = \frac{MTTR_{11}}{MTBF_{11}} = f_{11}MTTR_{11} = 0.022 \cdot 9.13 \cdot 10^{-4} = 2.01 \cdot 10^{-5} \quad (3.40)$$

This equals $8760 \cdot 2.01 \cdot 10^{-5} = 0.176 \text{h/y}$.
Alternatively, one might choose to write:

$$U_{11} = \frac{r_{11}}{T_{11}} = f_{11}r_{11} = 0.022 \cdot 9.13 \cdot 10^{-4} = 2.01 \cdot 10^{-5} \quad (3.41)$$

Or, if one takes into account that the repair time is measured in [h]:

$$U_{11} = \frac{r_{11h}}{T_{11} \cdot 8760} = f_{11}\frac{r_{11h}}{8760} = 0.022 \cdot \frac{8}{8760} = 2.01 \cdot 10^{-5} \quad (3.42)$$

Example 3.7: Failure Frequency Underground Cable

Question A specific underground cable circuit is on average 50 h/y unavailable. The repair time of this circuit is 730 h. What is the failure frequency of this cable circuit?

Answer The unavailability is given in [h/y], so it first must be converted to a (dimensionless) probability:

$$U_{c1} = \frac{50}{8760} = 5.71 \cdot 10^{-3} \quad (3.43)$$

The repair time is converted to the unit [y]:

$$MTTR_{c1} = \frac{730}{8760} = 8.33 \cdot 10^{-2}\,\text{y} \quad (3.44)$$

The failure frequency of the cable circuit can then be calculated as:

$$f_{c1} = \frac{1}{MTBF_{c1}} = \frac{MTTR_{c1}}{MTBF_{c1}MTTR_{c1}} = \frac{U_{c1}}{MTTR_{c1}} = \frac{5.71 \cdot 10^{-3}}{8.33 \cdot 10^{-2}} = 0.0685/y$$

(3.45)

Alternatively, one might choose to write:

$$f_{c1} = \frac{1}{T_{c1}} = \frac{r_{c1}}{T_{c1}r_{c1}} = \frac{U_{c1}}{r_{c1}} = \frac{5.71 \cdot 10^{-3}}{8.33 \cdot 10^{-2}} = 0.0685/y \qquad (3.46)$$

Or, if one takes into account that the repair time is measured in [h]:

$$f_{c1} = \frac{1}{T_{c1}} = \frac{r_{c1h} \cdot 8760}{T_{c1}r_{c1h} \cdot 8760} = \frac{U_{c1} \cdot 8760}{r_{c1h}} = \frac{5.71 \cdot 10^{-3} \cdot 8760}{730} = 0.0685/y$$

(3.47)

Example 3.8: Failure Rate and Failure Frequency 1

Question Generally, it is assumed that $f \approx \lambda$ for $MTTR \ll MTTF$. Assuming we have a 10 km overhead line circuit with failure frequency 0.022/y and repair time 8 h, what is the failure rate?

Answer The failure rate can be derived from the failure frequency and repair time:

$$\lambda_{11} = \frac{1}{MTTF_{11}} = \frac{1}{MTBF_{11} - MTTR_{11}} = \frac{1}{\frac{1}{f_{11}} - MTTR_{11}}$$

$$= \frac{1}{\frac{1}{0.022} - \frac{8}{8760}} = 0.022000442/y \qquad (3.48)$$

The result is (almost) equal to the failure frequency, as $MTTR_{11} \ll MTTF_{11}$.

Example 3.9: Failure Rate and Failure Frequency 2

Question Suppose we have a 10 km underground cable circuit with failure frequency 0.069/y. What will be the failure rate if the repair time is 168 h (1 week), 336 h (2 weeks) and 730 h (1 month)? What is the failure rate if the cable circuit is 100 km, with failure frequency 0.690/y? What is the increase in %?

Answer The failure rate can be calculated from the failure frequency and repair time as:

Table 3.3 Failure rate of a cable circuit for various repair times and lengths (Example 3.9)

Repair time	168 h	336 h	730 h
10 km circuit	0.069091 (= +0.1%)	0.069183 (= +0.3%)	0.069399 (= +0.6%)
100 km circuit	0.699253 (= +1.3%)	0.708758 (= +2.7%)	0.732095 (= +6.1%)

$$\lambda_{c1} = \frac{1}{MTTF_{c1}} = \frac{1}{MTBF_{c1} - MTTR_{c1}} = \frac{1}{\frac{1}{f_{c1}} - MTTR_{c1}} \qquad (3.49)$$

The increase in % is calculated as:

$$\frac{\lambda_{c1} - f_{c1}}{f_{c1}} \cdot 100\% \qquad (3.50)$$

The values then become as shown in Table 3.3. The error is very small and can be neglected for shorter cable circuits. For (very) long cable circuits and long repair times, it might be necessary to take the difference between failure frequency and failure rate into account in further reliability analysis.

Example 3.10: Failure Frequency Estimation

Question The failure frequency of components is often estimated from failure statistics. In the Netherlands, failure statistics of the components of the power grid are collected in the NESTOR database [14]. Suppose that failure statistics of overhead lines (OHLs) are collected for 5 years. 25 failures in EHV(380/220 kV) OHLs and 51 in HV(150/110 kV) OHLs were reported. The average total OHL circuit length (averaged over the 5 studied years) is 2310 km for the EHV and 3329 km for the HV network. Calculate the failure frequency (in /cctkm·y, per circuit-kilometer-year) for EHV OHLs, HV OHLs and (EHV/HV) OHLs in general, using the statistical sample mean (Appendix A.2). Which one is the largest?

Answer The failure frequencies can be calculated as:

$$f_{EHV} = \frac{F}{T} = \frac{25}{2310 \cdot 5} = 0.0022 \,/\text{cctkm·y} \qquad (3.51)$$

$$f_{HV} = \frac{F}{T} = \frac{51}{3329 \cdot 5} = 0.0031 \,/\text{cctkm·y} \qquad (3.52)$$

$$f_{OHL} = \frac{F}{T} = \frac{25 + 51}{(2310 + 3329) \cdot 5} = 0.0027 \,/\text{cctkm·y} \qquad (3.53)$$

The failure frequency of HV OHLs is the largest.

Question As more failure statistics will provide a higher accuracy, confidence intervals can be calculated to indicate the accuracy of the estimation. Confidence intervals of failure rates can be calculated by [11]:

$$\frac{\chi^2_{1-\alpha/2}(2F)}{2T} \leq \lambda \leq \frac{\chi^2_{\alpha/2}(2F+2)}{2T} \tag{3.54}$$

where:
λ = average failure rate [/cctkm·y] or [/comp·y]
T = total considered (component-)time length [cctkm·y] or [comp·y]
F = statistical number of failures within T [-]
α = significance level [-]
χ^2 = Chi-square distribution

Assume that $f \approx \lambda$ (because $MTTR \ll MTTF$). Calculate the 95%-confidence intervals ($\alpha = 0.05$) for the failure frequencies of OHV, HV and EHV/HV OHLs. This can for example be done using the function CHIINV in Microsoft Excel or with Matlab. What can be said about the difference between EHV and HV OHLs?

Answer For EHV OHLs, the confidence interval can be calculated as:

$$f_{EHVmin} = \frac{\chi^2_{1-0.05/2}(2 \cdot 25)}{2 \cdot 2310 \cdot 5} = 0.0014 / \text{cctkm·y} \tag{3.55}$$

$$f_{EHVmax} = \frac{\chi^2_{0.05/2}(2 \cdot 25 + 2)}{2 \cdot 2310 \cdot 5} = 0.0032 / \text{cctkm·y} \tag{3.56}$$

Or, in Microsoft Excel:
min EHV. CHIINV(0.975;2*25)/(2*2310*5) = 0.0014/cctkm·y
max EHV: CHIINV(0.025;2*25+2)/(2*2310*5) = 0.0032/cctkm·y

The other confidence intervals can be calculated similarly. The confidence intervals then become as shown in Table 3.4. The failure frequency of HV overhead lines is the largest, but there is a large overlap between the confidence intervals of EHV and HV OHLs, so it cannot be concluded that the failure frequency of HV OHLs is larger than the failures frequency of EHV OHLs.

Question Three years later, there are more failure statistics available. Suppose we have now 42 EHV OHL failures and 167 HV OHL failures. The average total OHL circuit lengths are 2471 km for the EHV network and 4078 km for the HV network. Calculate the failure frequencies of EHV, HV and EHV/HV OHLs again, together with their 95%-confidence intervals. What can be said about the difference between EHV and HV OHLs?

Table 3.4 Failure frequency and confidence levels of overhead lines (5 years) (Example 3.10)

Component	Lower confidence level	Average	Upper confidence level
EHV OHLs	0.0014	0.0022	0.0032
HV OHLs	0.0023	0.0031	0.0040
EHV/HV OHLs	0.0021	0.0027	0.0034

Table 3.5 Failure frequency and confidence levels of overhead lines (8 years) (Example 3.10)

Component	Lower confidence level	Average	Upper confidence level
EHV OHLs	0.0015	0.0021	0.0029
HV OHLs	0.0044	0.0051	0.0060
EHV/HV OHLs	0.0035	0.0040	0.0046

Answer If the calculation is performed again for the new values, the results become as shown in Table 3.5. The failure frequency of HV OHLs is the largest. As there is no overlap between the confidence intervals of EHV and HV OHLs, it can be concluded that the failure frequency of HV OHLs is larger than the failure frequency of EHV OHLs. Possible reasons could be that EHV OHLs are higher above the ground and have wider corridors than HV OHLs.

The same approach was followed to calculate the confidence intervals as shown in Fig. 2.1. In these graphs, it can be clearly seen that more failure statistics lead to more accurate estimations of the failure frequency.

3.3 Two-State Markov Model

Another model often used for the reliability behavior of components is the *Markov model* [1, 2]. A Markov model is a stochastic model based on the states in which a system can reside and the possible transitions between these states. The two stochastic variables *state S* and *time t* play an important role. These two variables can either be discrete or continuous, leading to four different kinds of Markov models. The case in which the states are discrete and time continuous is the most common in reliability analysis and is also called *Markov process*.

3.3.1 Unrepairable Component

As mentioned, a Markov model describes the states in which a system can be. These states are mutually exclusive, and the system always resides in one of these states. In

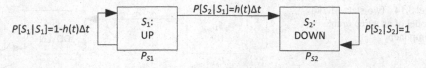

Fig. 3.13 Two-state Markov model of an unrepairable component [1]

its simplest form, the Markov model of a single (unrepairable) component consists of an UP state and a DOWN state, like shown in Fig. 3.13. The corresponding states can be called S_1 and S_2. The probabilities of these states are indicated by P_{S_1} and P_{S_2}.

Apart from the system states, a Markov model also shows the possible state transitions. Figure 3.13 shows the state transition from state S_1 to S_2, which is the failure process of the component. Furthermore, the probabilities that the component stays in state S_1 or S_2 are indicated in the figure. In Fig. 3.13, it is still assumed that the Markov model is time discrete.

In a *single-order Markov model*, the state transitions to other states only depend on the state in which the system currently is, and not on the previous states [2]. The state transitions therefore are conditional probabilities, and the probability of the transition from state S_i to state S_j can be described by [1, 2]:

$$P_{ij}(t, \Delta t) = P[S(t + \Delta t) = j | S(t) = i] = h_{ij}(t)\Delta t \tag{3.57}$$

In this equation, h_{ij} is called the *(state) transition rate* between states S_i and S_j, and is defined as [2]:

$$h_{ij}(t) = \lim_{\Delta t \to 0} \frac{P_{ij}(t, \Delta t)}{\Delta t} \tag{3.58}$$

For the two-state Markov model of Fig. 3.13:

$$P_{ij}(t, \Delta t) = h_{ij}(t)\Delta t = P[t < t_f \le t + \Delta t | t_f > t]$$
$$= \frac{P[t < t_f \le t + \Delta t]}{P[t_f > t]} = \frac{f(t)}{R(t)}\Delta t \tag{3.59}$$

The probability that the system is in a particular state thus depends on the probability that the system was in each state, the transition rates, and time. For the Markov model of Fig. 3.13:

$$\left[P[S(t + \Delta t) = 1] \ P[S(t + \Delta t) = 2]] \right] =$$
$$= \left[P[S(t) = 1] \ P[S(t) = 2]] \right] \cdot \begin{bmatrix} P[S_1|S_1] \ P[S_2|S_1] \\ P[S_1|S_2] \ P[S_2|S_2] \end{bmatrix} \tag{3.60}$$

Fig. 3.14 Two-state Markov model of an unrepairable component

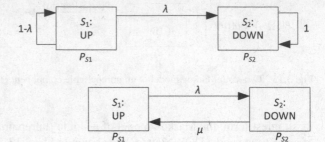

Fig. 3.15 Two-state Markov model of a repairable component

$$\big[P[S(t + \Delta t) = 1]\ P[S(t + \Delta t) = 2]\big] =$$

$$= \big[P[S(t) = 1]\ P[S(t) = 2]\big] \cdot \begin{bmatrix} 1 - h(t)\Delta t & h(t)\Delta t \\ 0 & 1 \end{bmatrix} \quad (3.61)$$

$$\mathbf{p}(t + \Delta t) = \mathbf{p}(t)T(\Delta t) \quad (3.62)$$

where $\mathbf{p}$ is called the *state probability vector* and T is the *transition matrix*.

If the transition rates in a Markov model are independent of the time, the Markov model is called *homogeneous* and the transition rate becomes the failure rate (λ) for the two-state Markov model as described before. Furthermore, the state probabilities then follow the (negative) exponential distribution [2]. The Markov model of Fig. 3.13 then becomes as shown in Fig. 3.14, while the transition matrix can be written as:

$$T = \begin{bmatrix} 1 - \lambda & \lambda \\ 0 & 1 \end{bmatrix} \quad (3.63)$$

3.3.2 Repairable Component

For a repairable component, the repair process is included in the two-state Markov model as well. As shown in Fig. 3.15, this is the state transition from state S_2 to state S_1. The state transitions are now indicated with the failure rate (λ) and repair rate (μ), respectively. State transitions from and to the same state are often omitted from the model.

The transition matrix now becomes:

$$T = \begin{bmatrix} 1 - \lambda & \lambda \\ \mu & 1 - \mu \end{bmatrix} \quad (3.64)$$

Although Markov models can describe the time behavior of systems (e.g. a new component is more likely to be in the UP state, while it is more likely to be in the DOWN state as time increases), Markov models are often assumed to be in *equilibrium* in reliability analysis. In equilibrium, the state probabilities do not change with time anymore and have thereby become time-independent.

For the transition matrix, this means [11]:

$$\mathbf{p}T = \mathbf{p} \tag{3.65}$$

$$\mathbf{p}(T - I) = \mathbf{0} \tag{3.66}$$

where I is the identity matrix. This will be used later when discussing larger Markov models in Sect. 4.2. Equilibrium normally occurs when $t \to \infty$.

From the state transitions in a Markov model, the *(state) transition frequencies* describe how often these state transitions occur. The (state) transition frequencies can be calculated as:

$$f_{S_i \to S_j} = P_{S_i} h_{ij} \tag{3.67}$$

Equilibrium means that the sum of the (state) transition frequencies entering a state must be equal to the sum of the state transition frequencies leaving a state:

$$\sum f_{\text{in}} = \sum f_{\text{out}} \tag{3.68}$$

This is logical because if these are not equal, the state probabilities will increase/decrease over time. In other words, a component must be repaired as often as it fails to be in equilibrium. For example, the state transition frequencies of the Markov model in Fig. 3.15 are:

$$f_{S_1 \to S_2} = P_{S_1} \lambda \tag{3.69}$$

$$f_{S_2 \to S_1} = P_{S_2} \mu \tag{3.70}$$

For state S_2 in Fig. 3.15, (3.68) becomes:

$$f_{S_1 \to S_2} = f_{S_2 \to S_1} \tag{3.71}$$

$$P_{S_1} \lambda = P_{S_2} \mu \tag{3.72}$$

Furthermore, as the system must be in one of the Markov states:

$$\sum P_{S_i} = 1 \tag{3.73}$$

Such that for Fig. 3.15:

$$P_{S_1} + P_{S_2} = 1 \tag{3.74}$$

$$P_{S_2} = 1 - P_{S_1} \tag{3.75}$$

Consequently:

$$P_{S_1}\lambda = P_{S_2}\mu = \left(1 - P_{S_1}\right)\mu = \mu - P_{S_1}\mu \tag{3.76}$$

$$P_{S_1}\lambda + P_{S_1}\mu = P_{S_1}\left(\lambda + \mu\right) = \mu \tag{3.77}$$

$$P_{S_1} = \frac{\mu}{\lambda + \mu} = A \tag{3.78}$$

The probability of state S_1 (UP) equals the availability A (according to (3.37)). Similarly, the probability of state S_2 (DOWN) equals the unavailability of the component.

Example 3.11: Markov State Probability

Question Show that using the previously described approach, the probability of state S_2 (DOWN) becomes the unavailability of the component in the two-state Markov model.

Answer For state S_1:

$$\sum f_{in} = \sum f_{out} \tag{3.79}$$

$$f_{S_2 \to S_1} = f_{S_1 \to S_2} \tag{3.80}$$

$$P_{S_2}\mu = P_{S_1}\lambda \tag{3.81}$$

Because $P_{S_1} + P_{S_2} = 1$:

$$P_{S_2}\mu = \left(1 - P_{S_2}\right)\lambda = \lambda - P_{S_2}\lambda \tag{3.82}$$

$$P_{S_2}\mu + P_{S_2}\lambda = P_{S_2}\left(\mu + \lambda\right) = \lambda \tag{3.83}$$

$$P_{S_2} = \frac{\lambda}{\mu + \lambda} = U \tag{3.84}$$

Example 3.12: Markov Model HVDC Converter Station

Question An HVDC converter station has a failure frequency of $f_{conv} = 2/y$ and a repair time of $r_{conv} = 24\,h$. Draw the two-state Markov model of this converter station and indicate the failure rate and repair rate in the model. What are the state probabilities? How many hours/year is the converter unavailable?

Fig. 3.16 The two-state Markov model of the converter station (Example 3.12)

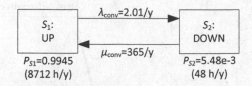

Answer The failure rate can be calculated from the failure frequency and repair time by (see Example 3.8):

$$\lambda_{\text{conv}} = \frac{1}{MTTF_{\text{conv}}} = \frac{1}{MTBF_{\text{conv}} - MTTR_{\text{conv}}}$$

$$= \frac{1}{\frac{1}{f_{\text{conv}}} - MTTR_{\text{conv}}} = \frac{1}{\frac{1}{2} - \frac{24}{8760}} = 2.01/y \qquad (3.85)$$

The repair rate can be calculated by:

$$\mu_{\text{conv}} = \frac{1}{MTTR_{\text{conv}}} = \frac{1}{24/8760} = 365/y \qquad (3.86)$$

(the component could be repaired $365\times$ per year if it were always broken)

The state probabilities of the two-state Markov model can now be calculated as:

$$P_{S_1} = \frac{\mu}{\lambda + \mu} = \frac{365}{2.01 + 365} = 0.9945 \qquad (3.87)$$

$$P_{S_2} = \frac{\lambda}{\mu + \lambda} = \frac{2.01}{365 + 2.01} = 5.48 \cdot 10^{-3} \qquad (3.88)$$

The converter is $8760 \cdot P_{S_2} = 8760 \cdot 5.48 \cdot 10^{-3} = 48\,h/y$ unavailable. This is in line with the failure frequency of $f_{\text{conv}} = 2/y$ and the repair time of $r_{\text{conv}} = 24\,h$. The two-state Markov model then becomes as shown in Fig. 3.16.

3.4 Stress-Strength Model

The last model that is discussed in this chapter is the *stress-strength model*. The stress-strength model assumes that both the *system strength* and the *system stress* can be described by probability distributions, as illustrated in Fig. 3.17. If the stress on the component is larger than the strength of the component, the component fails. This is the area in the graph where both probability distributions overlap.

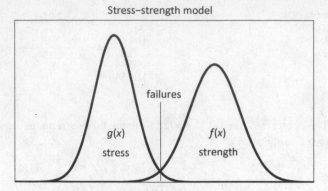

Fig. 3.17 Stress-strength model

If the system strength is described by probability distribution $f_{\text{strength}}(x)$, and the system stress by probability distribution $g_{\text{stress}}(x)$, the probability of failure can be calculated as follows [1, 5, 16]:

$$P[\text{strength} < \text{stress}] = \int_0^\infty g_{\text{stress}}(x) \int_0^x f_{\text{strength}}(y)\mathrm{d}y\mathrm{d}x$$

$$= \int_0^\infty g_{\text{stress}}(x) F_{\text{strength}}(x)\mathrm{d}x \qquad (3.89)$$

Although the stress-strength model can be used to model the reliability of individual components, it is also used to study the reliability of the generation system. Then, it is studied whether all generators of a power system together are strong enough to supply the load. This will be described in more detail in Sect. 5.2.

3.5 Conclusion

In this chapter, the reliability modeling of components was discussed. First, it was discussed how the failure behavior of unrepairable components can be described by reliability functions. It was shown how these reliability functions can be determined based on data analysis. Some important probability distributions (e.g. negative exponential and Weibull) and the bathtub curve were presented, as these are often used to model the failure behavior of components. For repairable components, the component life cycle was described and it was shown how a two-state Markov model can be used to model component reliability behavior. The stress-strength model was shortly introduced.

Reliability models of components are used for reliability studies of both small and large networks. The choice for a certain model always depends on the particular kind of study. For example, for asset management it is important to accurately model the reliability of single components. Data analysis, failure prediction, and maintenance planning are important topics in asset management. For reliability analysis of large networks, a simple modeling of components is beneficial, because using complex component models would lead to unmanageable large system models. Therefore, simplified component models like the exponential distribution and the component life cycle are preferred.

Problems

3.1 Learning Objectives
Please check the learning objectives as given at the beginning of this chapter.

3.2 Main Concepts
Please check whether you understand the following concepts:

- Repairable/unrepairable components
- Reliability functions:
  - Unreliability function or failure distribution ($F(t)$);
  - Reliability function $R(t)$;
  - Failure density distribution $f(t)$;
  - Hazard rate $h(t)$, failure rate λ.
- Probability distributions:
  - Normal distribution;
  - (Negative) exponential distribution;
  - Weibull distribution.
- Data analysis:
  - Uncensored and censored data;
  - Left and right censored data;
  - Accelerated testing;
  - Weibull analysis.
- Bathtub curve:
  - Infant stage;
  - Normal operating stage;
  - Wear-out stage;
  - Time-based and condition-based maintenance.
- Component life cycle:
  - Mean time to failure ($MTTF$ or d);
  - Mean time to repair or repair time ($MTTR$ or r);
  - Mean time between failures ($MTBF$ or T);

- Failure rate (λ) / Repair rate (μ);
- Failure frequency (f);
- Availability (A);
- Unavailability (U).

- Markov models:

  - Markov model and Markov process;
  - Homogenous Markov model;
  - Single-order Markov model;
  - System state (S) and time (t);
  - (State) transition rate and frequency;
  - Transition matrix and state probability vector;
  - Equilibrium.

- Stress-strength model:

  - System strength;
  - System stress.

3.3 Reliability Functions Component

Suppose we study the failures of ten components to derive reliability functions of this particular kind of component. We found that these components failed after 2, 4, 6, 9, 12, 16, 20, 24, 30, and 36 years, respectively. Based on this information, draw the reliability functions of this component.

3.4 Reliability Functions

What are the four reliability functions? How are they (mathematically) related?

3.5 Probability Distributions

Describe three probability distributions used to model the failure behavior of components.

3.6 Data Analysis

Suppose we have six components:

- Component A was installed at time $t = 0$ and failed after 4 years.
- Component B was installed at time $t = 0$ and failed after 6 years.
- Component C was installed at time $t = 3$ and failed after 8 years.
- Component D was installed at time $t = 4$ and survived the 10 years observation time.
- Component E was installed at time $t = -1$ and was removed from observation (without failure) at $t = 8$.
- Component F was installed before $t = 0$ and failed after 4 years.

Describe how this information is used in data analysis to derive the reliability functions. Clearly describe what (uncensored/censored) data and (left/right) censored data is.

3.7 Bathtub Curve

Draw the bathtub curve and indicate the three stages of the component lifetime. Indicate how testing, (time-based/condition-based) maintenance, and replacement are used to improve the component reliability and increase its lifetime.

3.8 Negative Exponential Distribution

Negative exponential distribution:

- What is the (negative) exponential distribution?
- What is the main assumption of this distribution?
- What is the reliability function ($R(t)$) according to the exponential distribution?
- Derive the other three reliability functions from $R(t)$.
- Sketch the graphs of these reliability functions.

3.9 Weibull Analysis

Describe (in words/sketches) what Weibull analysis is and how it is used for reliability analysis of components.

3.10 Component Life Cycle

Draw the life cycle of a component. Try to derive various related functions, for example, the failure frequency (f) as a function of mean time to failure ($MTTF$) and mean time to repair ($MTTR$). Or the failure rate (λ) as a function of the failure frequency (f) and the repair time (r).

3.11 Failure Rate Transformer

An offshore transformer has a failure frequency of $f_{tr} = 0.05/y$ and a repair time of $r_{tr} = 1460\,h$. What is its unavailability (U_{tr})? What is its failure rate (λ_{tr})?

3.12 Repair Time Overhead Line

An overhead line circuit is on average 0.1 h/y unavailable and has a failure frequency of $f_{ohl} = 0.1/y$. What is its repair time (in [h])?

3.13 Failure Frequencies & Repair Times

Study the figures and tables of the failure frequencies and repair times of components in Chap. 2 again. What can you conclude?

3.14 Markov Model Offshore Cable

A 10 km offshore HVDC cable has a failure frequency of $f_{oc} = 0.70/y$ and a repair time of $r_{oc} = 730\,h$. Draw a (two-state) Markov model for this cable. Clearly indicate all the parameters in the diagram. What are the state probabilities? What is the main condition to solve the Markov model in equilibrium?

3.15 Markov Model Generator

A generator has a failure frequency of $f_{gen} = 5/y$ and a repair time of $r_{gen} = 48\,h$. Draw the two-state Markov model of this generator and indicate the failure rate and repair rate in the model. What are the state probabilities? How many hours/year is the generator unavailable on average?

3.16 Stress-Strength Model

Describe (in words/sketches) how the stress-strength model is used for reliability analysis of components.

References

1. Klaassen KB, van Peppen JCL, Bossche A (1988) Bedrijfszekerheid, Theorie en Techniek. Delftse Uitgevers Maatschappij, Delft, The Netherlands
2. Kling WL (2006) Planning en Bedrijfsvoering van Electriciteitsvoorzieningssystemen, lecture notes. Delft University of Technology, Delft, The Netherlands
3. Yates RD, Goodman DJ (2005) Probability and stochastic processes. Wiley, United States
4. Billinton R, Allan RN (1992) Reliability evaluation of engineering systems, 2nd edn. Plenum Press, New York
5. Bossche A (2016) Microelectronics reliability - lecture notes. Delft University of Technology, Delft, The Netherlands
6. Zhuang Q (2015) Managing risks in electrical infrastructure assets from a strategic perspective. PhD thesis, Delft University of Technology, Delft, The Netherlands
7. Ross R (2013) Betrouwbaarheid en Prestaties bij Asset Management van Systemen. IWO - Instituut voor Wetenschap en Duurzame Ontwikkeling, The Netherlands
8. Kreuger FH (1992) Industrial high voltage - co-ordinating, measuring, testing. Delft University Press, Delft, The Netherlands,
9. Meijer S, Smit J, Chen X, Fischer W, Colla L (2011) Return of experience of 380 kV XLPE landcable failures. In: Jicable - 8th International conference on insulated power cables, Versailles, France
10. Meijer S, Smit J, Chen X, Gulski E (2011) Monitoring facilities for failure rate reduction of 380 kV power cables. In: Jicable - 8th international conference on insulated power cables, Versailles, France
11. Li W (2005) Risk assessment of power systems - models, methods, and applications. Wiley Interscience - IEEE Press, Canada
12. Mathworks (2011) MATLAB product description. www.mathworks.com/products/matlab
13. VDN (2004) Ermittlung von Eingangsdaten für Zuverlässigkeitsberechnungen aus der VDN-Störungsstatistik. VDN, Germany. http://www.vde.com/de/fnn/arbeitsgebiete/versorgungsqualitaet/seiten/versorgungszuverlaessigkeit-unterlagen.aspx
14. TenneT TSO (2011) NESTOR (Nederlandse Storingsregistratie) database 2006–2010. TenneT TSO B.V, Arnhem, Netherlands
15. Billinton R, Allan RN (1984) Power-system reliability in perspective. Electronics and power, vol 30, pp 231–236
16. Wikipedia (2016) Stress-strength analysis. http://reliawiki.org/index.php/Stress-StrengthAnalysis

Chapter 4
Reliability Models of Small Systems

The previous chapter described how to model the reliability of individual components. By combining several individual components, small systems are created. The reliability of these systems can be analyzed by combining the reliability models of the individual components. This chapter discusses the methods that are often used for reliability analysis of small systems. This chapter starts with a description of reliability networks in Sect. 4.1. Then, the application of Markov models to small systems is discussed in Sect. 4.2. Section 4.3 addresses how fault tree and event tree analysis can be used for reliability calculations. Final conclusions are then discussed in Sect. 4.4.

> **Learning Objectives**
> After reading this chapter, the student should be able to:
> - describe which methods are used for reliability analysis of small systems;
> - create and solve reliability networks of small systems;
> - create and solve Markov models of small systems;
> - apply fault tree and event tree analysis.

4.1 Reliability Networks

Section 3.2 discussed how the reliability of components can be described by their availability (A) and unavailability (U). When individual components are connected, small networks are created. These networks can be modeled as *reliability networks* (i.e. series/parallel networks) in reliability analysis. One must be aware here of the fact that reliability networks in reliability analysis do not necessarily need to be similar to the network configuration in reality. For example, when all components of

© Springer Nature Switzerland AG 2020
B. W. Tuinema et al., *Probabilistic Reliability Analysis of Power Systems*,
https://doi.org/10.1007/978-3-030-43498-4_4

Fig. 4.1 Series connection

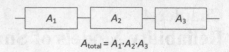

$$A_{\text{total}} = A_1 {\cdot} A_2 {\cdot} A_3$$

a parallel network are needed for correct operation, the network is a series connection in reliability analysis. This section will show that the reliability of reliability networks can often be calculated easily.

4.1.1 Series Connections

A *series connection* is created by connecting multiple components in series, as illustrated in Fig. 4.1. The connection is available if all of the individual components are available, such that the availability of the connection becomes the product of the availabilities of the components [1, 2].

$$A_{\text{total}} = \prod_{i=1}^{N} A_i \tag{4.1}$$

where:
A_{total} = availability of the connection
N = number of components in the connection
A_i = availability of component i

Also, the reliability function of the connection is the product of the reliability functions of the components:

$$R_{\text{total}}(t) = \prod_{i=1}^{N} R_i(t) \tag{4.2}$$

where:
R_{total} = reliability function of the connection
R_i = reliability function of component i

And if the reliability functions are described by exponential distributions, the failure rate of the connection is the sum of the failure rates of the components:

$$R_{\text{total}}(t) = \prod_{i=1}^{N} R_i(t) = \prod_{i=1}^{N} e^{-\lambda_i t} = e^{-\sum_{i=1}^{N} \lambda_i t} \tag{4.3}$$

$$\lambda_{\text{total}} = \sum_{i=1}^{N} \lambda_i \tag{4.4}$$

where:
λ_{total} = failure rate of the connection
λ_i = failure rate of component i

Example 4.1: Series Connection Offshore Wind

Question An offshore wind farm is connected to the shore by a series connection consisting of an offshore transformer (with failure frequency $f_{\text{tr1}} = 0.060/\text{y}$ and repair time $r_{\text{tr1}} = 500\,\text{h}$), a 10 km submarine cable (with failure frequency $f_{\text{c1}} = 0.010/\text{y}$ and repair time $r_{\text{c1}} = 1400\,\text{h}$), and an onshore transformer (with failure frequency $f_{\text{tr2}} = 0.040/\text{y}$ and repair time $r_{\text{tr2}} = 48\,\text{h}$). The configuration of this connection is illustrated in Fig. 4.2. It is assumed that the offshore wind farm and onshore substation are fully reliable. What is the unavailability of the (transformer-cable-transformer) connection? How many hours/year is the connection unavailable on average?

Answer Combining the theory of this section with the theory discussed in Sect. 3.2, the unavailability of the connection can be calculated as follows:

$$U_{\text{con1}} = 1 - A_{\text{con1}} = 1 - (A_{\text{tr1}} A_{\text{c1}} A_{\text{tr2}}) = 1 - ((1 - U_{\text{tr1}})(1 - U_{\text{c1}})(1 - U_{\text{tr2}}))$$

$$= 1 - \left(\left(1 - \frac{f_{\text{tr1}} r_{\text{tr1}}}{8760} \right) \left(1 - \frac{f_{\text{c1}} r_{\text{c1}}}{8760} \right) \left(1 - \frac{f_{\text{tr2}} r_{\text{tr2}}}{8760} \right) \right) \qquad (4.5)$$

$$= 1 - \left(\left(1 - \frac{0.060 \cdot 500}{8760} \right) \left(1 - \frac{0.010 \cdot 1400}{8760} \right) \left(1 - \frac{0.040 \cdot 48}{8760} \right) \right) = 5.24 \cdot 10^{-3}$$

It is expected that this connection is unavailable $5.24 \cdot 10^{-3} \cdot 8760 = 45.9\,\text{h/y}$.

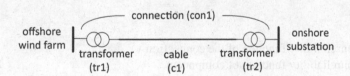

offshore wind farm — connection (con1) — onshore substation

transformer (tr1) cable (c1) transformer (tr2)

Fig. 4.2 Connection of an offshore wind farm by one submarine cable (Example 4.1)

Fig. 4.3 Parallel connection
(1-out-of-N)

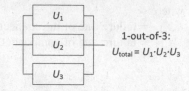

1-out-of-3:
$U_{\text{total}} = U_1{\cdot}U_2{\cdot}U_3$

4.1.2 Parallel Connections

A *parallel connection* is created by connecting multiple components in parallel, as illustrated in Fig. 4.3. The availability of the connection now depends on the number of components that is minimally required to have a working connection.

1-out-of-N
If at least one component is required, it is called a *1-out-of-N* connection. The connection is then unavailable if all of the individual components are unavailable, such that the unavailability of the connection becomes the product of the unavailabilities of the components [1, 2].

$$U_{\text{total}} = \prod_{i=1}^{N} U_i \tag{4.6}$$

where:
U_{total} = unavailability of the connection
N = number of components in the connection
U_i = unavailability of component i

Also, the unreliability function of the connection is the product of the unreliability functions of the components:

$$F_{\text{total}}(t) = \prod_{i=1}^{N} F_i(t) \tag{4.7}$$

where:
F_{total} = unreliability function of the connection
F_i = unreliability function of component i

N-out-of-N
If all components of a parallel network need to be available for a working connection, the connection is unavailable if any of the components is unavailable. This is effectively the same as a series connection of the components.

M-out-of-N
If at least m components of a parallel connection of N components are needed for a working connection, this is called an *m-out-of-N* connection [1]. If all components

have the same availability A_0, the availability of the total connection can then be calculated by:

$$A_{\text{total}} = \sum_{i=m}^{N} \binom{N}{i} (A_0)^i (1 - A_0)^{(N-i)} \tag{4.8}$$

Also, the reliability function of the total connection is:

$$R_{\text{total}}(t) = \sum_{i=m}^{N} \binom{N}{i} (R_0)^i (t) (1 - R_0(t))^{(N-i)} \tag{4.9}$$

In practice however, these m-out-of-N networks are somewhat more complicated. The equations above assume that the connection is either available or unavailable (as $A = 1 - U$) and the connection is either reliable or unreliable (as $R(t) = 1 - F(t)$). In reality, if there are two connections in parallel and one of these fails, still half the transmission capacity is available. Therefore, in practical studies, often the probabilities of having a certain transmission capacity are calculated. This is illustrated in the following two examples.

Example 4.2: Parallel Connection Offshore Wind

Question The offshore wind farm of Example 4.1 is now connected by two connections in parallel, as illustrated in Fig. 4.4. What is the probability of having 0/1/2 connections available in the offshore network? What is the probability that there is no transmission capacity if the offshore network is $N - 1$ redundant?

Answer In Example 4.1, it was calculated that $U_{\text{con1}} = 5.24 \cdot 10^{-3}$. As both connections consist of the same components:

$$U_{\text{con2}} = U_{\text{con1}} \tag{4.10}$$

$$A_{\text{con2}} = A_{\text{con1}} = 1 - U_{\text{con1}} = 1 - 5.24 \cdot 10^{-3} \tag{4.11}$$

The probabilities of having connection outages and having a certain transmission capacity then become as shown in Table 4.1. The probabilities of having 0/1/2 connections (i.e. 0/50/100% transmission capacity) available are thus respectively:

$$P_{0\%} = P_3 = 2.75 \cdot 10^{-5} \tag{4.12}$$

$$P_{50\%} = P_1 + P_2 = 2 \cdot 5.21 \cdot 10^{-3} = 1.04 \cdot 10^{-2} \tag{4.13}$$

$$P_{100\%} = P_0 = 0.9895 \tag{4.14}$$

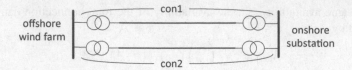

Fig. 4.4 Connection of an offshore wind farm by two submarine cables (Example 4.2)

Table 4.1 Probability of having a certain transmission capacity of the offshore network (Example 4.2)

Con1	Con2	Capacity(%)	Probability [−]
in	in	100	$P_0 = A_{con1} \cdot A_{con2} = 0.9895$
in	out	50	$P_1 = A_{con1} \cdot U_{con2} = 5.21 \cdot 10^{-3}$
out	in	50	$P_2 = U_{con1} \cdot A_{con2} = 5.21 \cdot 10^{-3}$
out	out	0	$P_3 = U_{con1} \cdot U_{con2} = 2.75 \cdot 10^{-5}$

If the offshore network is $N - 1$ redundant, a failure of one component will not lead to a failure of the complete connection. The offshore network only fails when both connections (con1 and con2) fail:

$$P_3 = 2.75 \cdot 10^{-5} \tag{4.15}$$

This equals $2.75 \cdot 10^{-5} \cdot 8760 = 0.24 \, \text{h/y}$.

Example 4.3: Parallel Connection Offshore Wind 2

Question The offshore wind farm of Example 4.1 is now connected by three connections in parallel, as illustrated in Fig. 4.5. What is the probability of having 100%/67%/33%/0% transmission capacity available?

Answer In Example 4.1, it was calculated that $U_{con1} = 5.24 \cdot 10^{-3}$. As all three (transformer-cable-transformer) connections consist of the same components:

$$U_{con3} = U_{con2} = U_{con1} \tag{4.16}$$

$$A_{con3} = A_{con2} = A_{con1} = 1 - U_{con1} = 1 - 5.24 \cdot 10^{-3} \tag{4.17}$$

The probabilities of having a certain transmission capacity then can be calculated by:

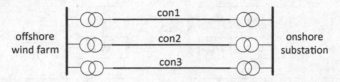

Fig. 4.5 Connection of an offshore wind farm by three submarine cables (Example 4.3)

Table 4.2 Probability of having a certain transmission capacity of the offshore network (Example 4.3)

Capacity [%]	Probability [−]	Expectation [h/y]
100	0.984	8623
67	$1.55 \cdot 10^{-2}$	136
33	$8.18 \cdot 10^{-5}$	0.717
0	$1.44 \cdot 10^{-7}$	$1.26 \cdot 10^{-3}$

$$P_{100\%} = \binom{3}{3} A_{\text{con1}}^3 = (1 - U_{\text{con1}})^3 = 0.984 \tag{4.18}$$

$$P_{67\%} = \binom{3}{2} A_{\text{con1}}^2 U_{\text{con1}} = 3(1 - U_{\text{con1}})^2 U_{\text{con1}} = 1.55 \cdot 10^{-2} \tag{4.19}$$

$$P_{33\%} = \binom{3}{1} A_{\text{con1}} U_{\text{con1}}^2 = 3(1 - U_{\text{con1}}) U_{\text{con1}}^2 = 8.18 \cdot 10^{-5} \tag{4.20}$$

$$P_{0\%} = \binom{3}{0} U_{\text{con1}}^3 = U_{\text{con1}}^3 = 1.44 \cdot 10^{-7} \tag{4.21}$$

The results are summarized in Table 4.2. It can be verified that the probabilities add up to 1.

4.1.3 Dependent Failures

As already mentioned before in Sect. 2.2.5, components can also fail dependently. Dependent (common-cause) failures can be included in reliability networks as well. Mostly, these common-cause failures are modeled as a separate component which is placed in series with the components that can fail dependently [1]. This is illustrated in Fig. 4.6.

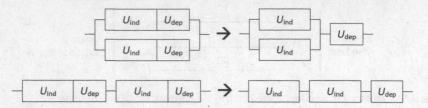

Fig. 4.6 Modeling of dependent (common-cause) failures

4.1.4 Mixed Series-Parallel Networks

For larger systems, modeling by reliability networks often results in mixed series-parallel networks. Sometimes, the reliability network can be solved in parts, like the network shown in Fig. 4.7. In other cases, the network can be solved by advanced techniques like *paths-cuts methods* or the *decomposition method* [1].

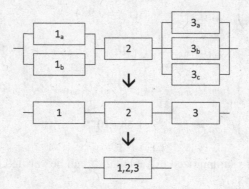

Fig. 4.7 Stepwise solution of mixed series-parallel network

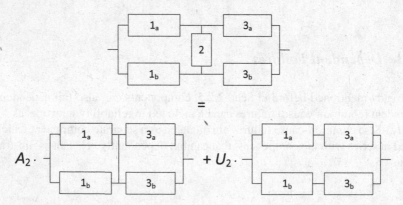

Fig. 4.8 Decomposition method

Decomposition Method

Consider the network shown in Fig. 4.8. This network cannot be solved by the series or parallel calculation rules alone. The critical component that makes this network complicated is component 2. If component 2 is available, the network can be solved easily. The same holds if component 2 is unavailable. Therefore, as illustrated in Fig. 4.8, the network can be decomposed into two networks by the *decomposition method*. The reliability of the connection then becomes the weighted sum of the reliability of the two networks, weighted to the (un)availability of component 2.

Example 4.4: Series-Parallel Connection Offshore Wind

Question The offshore wind farm of Example 4.1 is now connected to the offshore network as shown in Fig. 4.9. The offshore network is $N - 1$ redundant, such that with one failure in the offshore network, there is still enough capacity to transport the generated power to the shore. What is the probability that there is not enough transmission capacity in the offshore network?

Answer This offshore network topology has the same structure as the network in Fig. 4.8. To solve this network, the decomposition method can be applied to create two networks as shown in Fig. 4.10. Using the calculation rules for series and parallel networks:

$$U_1 = 1 - A_1 = 1 - A_{tr1} A_{cbl} = 1 - (1 - U_{tr1})(1 - U_{cbl}) = 5.0 \cdot 10^{-3}$$
(4.22)

$$U_2 = 1 - A_2 = 1 - A_{tr2} A_{cbl} = 1 - (1 - U_{tr2})(1 - U_{cbl}) = 2.2 \cdot 10^{-3}$$
(4.23)

The fact that the offshore network is $N - 1$ redundant is crucial when solving this network. If the offshore network is $N - 1$ redundant, the parallel connections become 1-out-of-N networks. Then, the network can be reduced according to Figs. 4.11 and 4.12.

$$U_3 = U_1^2 = 2.5 \cdot 10^{-5}$$
(4.24)

$$U_4 = U_2^2 = 4.8 \cdot 10^{-6}$$
(4.25)

$$U_5 = 1 - A_5 = 1 - A_1 A_2 = 1 - (1 - U_1)(1 - U_2) = 7.2 \cdot 10^{-3}$$
(4.26)

$$U_6 = 1 - A_6 = 1 - A_3 A_4 = 1 - (1 - U_3)(1 - U_4) = 3.0 \cdot 10^{-5}$$
(4.27)

$$U_7 = U_5^2 = 5.2 \cdot 10^{-5}$$
(4.28)

$$U_{total} = A_{br} U_6 + U_{br} U_7 = (1 - U_{br})U_6 + U_{br} U_7 = 3.0 \cdot 10^{-5}$$
(4.29)

It is expected that there is $8760 \cdot 3.0 \cdot 10^{-5} = 0.26 \, \text{h/y}$ not enough transmission capacity in the offshore network.

Because this particular offshore network is $N - 1$ redundant, there is either enough transmission capacity in the offshore network or there is no transmission capacity at all. Therefore, we can use the concept unavailability (as $U = 1 - A$) and calculate U_{total} by the decomposition method. If the offshore network is not $N - 1$ redundant, it is possible that there is, for example, 50% transmission capacity available in the offshore network because of a component failure. Then, it must be investigated how component failures affect the available transmission capacity. This could be done with methods like fault tree analysis (Sect. 4.3) or state enumeration (Sect. 5.1).

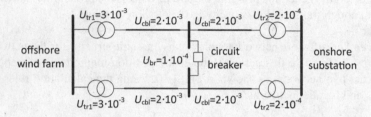

Fig. 4.9 Connection of an offshore wind farm by an offshore network (Example 4.4)

Fig. 4.10 Reliability network of the offshore network shown in Fig. 4.9 (Example 4.4)

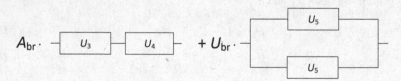

Fig. 4.11 Simplification of the reliability network of Fig. 4.10 (Example 4.4)

Fig. 4.12 Simplification of the reliability network of Fig. 4.11 (Example 4.4)

4.2 Markov Models

In Sect. 3.3, it was described how the failure behavior of individual components can be modeled by a two-state Markov model. Markov models can be expanded to include more system states. In this way, a Markov model can represent more system states of the same component. Moreover, the reliability of a (small) system of individual components can effectively be modeled by a Markov model. In this section, the general approach to create a Markov model and the application to individual components and small systems are described. Some techniques to reduce the size of a Markov model are discussed as well.

4.2.1 Creation of a Markov Model

A Markov model can be created by following these steps:

1. Start with a system state in which all components are working.
2. From this 'all-ok' state, investigate which component failures can occur.
3. For every possible component failure, draw a new system state. The transition rate to this new system state is the failure rate of the specific component.
4. For every new system state, investigate what failures (or repairs) might occur next:

   - If a failure (or repair) leads to a new system state, draw this system state. The transition rate to this new system state is the failure rate (or repair rate) of the specific component.
   - If a failure (or repair) leads to an already existing system state, the transition rate to this existing system state is the failure rate (or repair rate) of the specific component.

5. Repeat step 4 until there are no other new states or state transitions.
6. Double check whether all possible system states and state transitions are included in the Markov model.
7. If necessary and possible, apply reduction techniques to reduce the size of the Markov model.
8. Indicate the different categories of the system states in the Markov model, e.g. system is working or defect, system works at full/half/zero capacity.

The process of creating a Markov model will become more clear in the examples given in this section. Although the approach to create a Markov model is straightforward, the steps of the approach must be followed carefully. Mistakes are easily made as possible system states and state transitions are easily overlooked.

4.2.2 Solution of the Markov Model

Assuming that the Markov model is in equilibrium, it can now be solved in a similar way as described in Sect. 3.3. There are two approaches for this. In the first, the state transition frequencies can be calculated by (3.67):

$$f_{S_i \to S_j} = P_{S_i} h_{ij} \tag{4.30}$$

The equilibrium condition is then applied for each state (3.68):

$$\sum f_{\text{in}} = \sum f_{\text{out}} \tag{4.31}$$

As the examples in this section show, this can best be written as a matrix equation. Because this set of equations is linearly dependent, one equation must be replaced to include the condition that the system is in one of the Markov states (3.73):

$$\sum P_{S_i} = 1 \tag{4.32}$$

In the second approach, the transition matrix is created directly from the Markov model. The elements on the diagonal of this transition matrix are 1 minus the sum of the outgoing transition rates. The other positions in the matrix are the transition rates between the corresponding states (row i, column j: from state S_i to state S_j). The equilibrium condition is then applied (3.65):

$$\mathbf{p}T = \mathbf{p} \tag{4.33}$$

$$\mathbf{p}(T - I) = \mathbf{0} \tag{4.34}$$

If it is realized that this is the same as (using matrix transpose):

$$(T - I)^{\text{tr}}\mathbf{p}^{\text{tr}} = \mathbf{0}^{\text{tr}} \tag{4.35}$$

One finds that $(T - I)^{\text{tr}}$ is exactly the set of equations of the (state) transition frequencies using (4.31). This will become more clear in the examples in this section as well. Again, the condition of (4.32) must be included by replacing one row of this set of equations. Table 4.3 gives an overview of both approaches. From the second step, both approaches are similar.

4.2.3 Markov Models of Individual Components

The two-state Markov model described in Sect. 3.3 assumed that a component either is in the UP-state (working) or in the DOWN-state (defect). In reality, components

Table 4.3 Approaches to solve Markov models

Approach 1	Approach 2
Write down the (state) transition frequency equation for each state using: $\sum f_{in} = \sum f_{out}$.	Create the transition matrix T: the values on the diagonals are $1 - \sum h_{out}$; other positions are (row i, column j) $= h_{S_i \to S_j}$.
Write this in matrix notation.	Calculate $(T - I)^{tr}$.
Replace one equation to include $\sum P_{S_i} = 1$.	Replace one equation to include $\sum P_{S_i} = 1$.
Solve the resulting matrix equation.	Solve the resulting matrix equation.

can be in more states. For example, different failure modes of a component can lead to different states with different consequences. The two-state Markov model can then be expanded to include more possible states. An example is a generator which can fail partially. A partial failure of the generator then leads to a *derated state*, in which the generator produces less power than normal. A more critical failure can still lead to a complete shutdown of the generator.

Example 4.5: Markov Model Generator

Question A specific gas generator can fail in two ways. It can fail partially, with a failure rate of $\lambda_{g1} = 5/y$ and an average repair time of $r_{g1} = 24$ h. This leads to a derated state. It can also fail critically, with a failure rate of $\lambda_{g2} = 2/y$, which leads to a complete shutdown of the generator. If the generator is in the derated state, it can still fail critically. If the generator is completely shut down, it has an average repair time of $r_{g2} = 48$ h, after which it is in normal operation again. Draw the Markov model of this specific generator and indicate the failure rates and repair rates in the model. Solve the Markov model.

Answer Before creating the Markov model, the repair rates are calculated from the repair times:

$$\mu_{g1} = \frac{1}{r_{g1}/8760} = \frac{1}{24/8760} = 365/y \tag{4.36}$$

$$\mu_{g2} = \frac{1}{r_{g2}/8760} = \frac{1}{48/8760} = 182.5/y \tag{4.37}$$

To create the Markov model, the algorithm from Sect. 4.2.1 is followed. First, the normal operation state (S_1: UP) is drawn. From this state, the generator can fail partially, which leads to state S_2 (DERATED), with failure rate λ_{g1}. From state S_1, the generator can also fail critically, which leads to state S_3 (DOWN), with failure rate λ_{g2}. The result of steps 1 to 3 of the algorithm is shown in Fig. 4.13 on the left.

It is now investigated what can happen next when the system is in the created states. From state S_2, the generator can be repaired with repair rate μ_{g1}, which leads to state S_1. From state S_2, the generator can also fail critically with failure rate λ_{g2}, which leads to state S_3. From state S_3, the generator can be repaired with repair rate μ_{g2}, which leads to state S_1. The result of steps 4 to 8 of the algorithm are shown in Fig. 4.13 on the right. The created Markov model can be double checked to verify whether it corresponds to the given information. The three possible states are: UP, DERATED and DOWN.

The Markov model can be solved using Eqs. 4.30 and 4.31 (i.e. approach 1 of Table 4.3). For state S_1:

$$\sum f_{in} = \sum f_{out} \tag{4.38}$$

$$f_{S_2 \to S_1} + f_{S_3 \to S_1} = f_{S_1 \to S_2} + f_{S_1 \to S_3} \tag{4.39}$$

$$P_{S_2} \mu_{g1} + P_{S_3} \mu_{g2} = P_{S_1} \lambda_{g1} + P_{S_1} \lambda_{g2} \tag{4.40}$$

$$-(\lambda_{g1} + \lambda_{g2}) P_{S_1} + \mu_{g1} P_{S_2} + \mu_{g2} P_{S_3} = 0 \tag{4.41}$$

For state S_2:

$$f_{S_1 \to S_2} = f_{S_2 \to S_1} + f_{S_2 \to S_3} \tag{4.42}$$

$$P_{S_1} \lambda_{g1} = P_{S_2} \mu_{g1} + P_{S_2} \lambda_{g2} \tag{4.43}$$

$$\lambda_{g1} P_{S_1} - (\mu_{g1} + \lambda_{g2}) P_{S_2} = 0 \tag{4.44}$$

For state S_3:

$$f_{S_1 \to S_3} + f_{S_2 \to S_3} = f_{S_3 \to S_1} \tag{4.45}$$

$$P_{S_1} \lambda_{g2} + P_{S_2} \lambda_{g2} = P_{S_3} \mu_{g2} \tag{4.46}$$

$$\lambda_{g2} P_{S_1} + \lambda_{g2} P_{S_2} - \mu_{g2} P_{S_3} = 0 \tag{4.47}$$

Or, in matrix notation:

$$\begin{bmatrix} -(\lambda_{g1} + \lambda_{g2}) & \mu_{g1} & \mu_{g2} \\ \lambda_{g1} & -(\mu_{g1} + \lambda_{g2}) & 0 \\ \lambda_{g2} & \lambda_{g2} & -\mu_{g2} \end{bmatrix} \begin{bmatrix} P_{S_1} \\ P_{S_2} \\ P_{S_3} \end{bmatrix} = \begin{bmatrix} 0 \\ 0 \\ 0 \end{bmatrix} \tag{4.48}$$

It can be checked whether the matrix is a linearly dependent set of equations. The rows of the matrix equation add up to 0. If not, some mistake was made

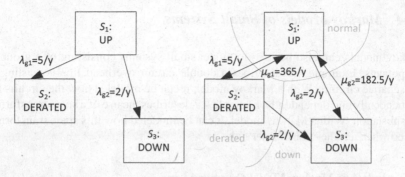

Fig. 4.13 Markov model of a generator (Example 4.5)

when writing down the equations. To solve this set of equations, one equation must be replaced by (4.32):

$$
\begin{bmatrix}
-(\lambda_{g1} + \lambda_{g2}) & \mu_{g1} & \mu_{g2} \\
\lambda_{g1} & -(\mu_{g1} + \lambda_{g2}) & 0 \\
1 & 1 & 1
\end{bmatrix}
\begin{bmatrix}
P_{S_1} \\
P_{S_2} \\
P_{S_3}
\end{bmatrix}
=
\begin{bmatrix}
0 \\
0 \\
1
\end{bmatrix}
\tag{4.49}
$$

This matrix equation can now be solved easily, e.g. by Matlab:

$$
\begin{bmatrix}
P_{S_1} \\
P_{S_2} \\
P_{S_3}
\end{bmatrix}
=
\begin{bmatrix}
0.9759 \\
0.0133 \\
0.0108
\end{bmatrix}
\tag{4.50}
$$

Consequently, it is expected that this generator is 8549 h/y in normal operation, 116 h/y in the derated state, and 95 h/y shut down completely.

One could also choose to write down the transition matrix directly from the Markov model (i.e. approach 2 of Table 4.3). The values on the diagonal are 1 minus the sum of the outgoing transition rates, while the other positions are the transition rates of the corresponding states:

$$
T =
\begin{bmatrix}
1 - (\lambda_{g1} + \lambda_{g2}) & \lambda_{g1} & \lambda_{g2} \\
\mu_{g1} & 1 - (\mu_{g1} + \lambda_{g2}) & \lambda_{g2} \\
\mu_{g2} & 0 & -\mu_{g2}
\end{bmatrix}
\tag{4.51}
$$

One can easily verify that $(T - I)^{\text{tr}}$ is the same matrix as above. Also, the further solution of the Markov model using the transition matrix is similar to the solution as described above.

4.2.4   Markov Models of Small Systems

Markov models can also be used to model small systems consisting of a group of components. Example 4.6 describes a double circuit overhead line consisting of two separate circuits. With a Markov model, it can be modeled how the circuits fail independently and dependently. Example 4.7 describes the use of a spare transformer in a substation. With a Markov model, it can be modeled how this spare transformer is used when another transformer fails.

Example 4.6: Markov Model Overhead Line

Question Assume that we have a double circuit overhead line connection of 10 km. The failure rate of one circuit is $\lambda_{11} = 0.022/y$, while the average repair time is $r_{11} = 8$ h. The circuits of the connection can fail independently. Both circuits of the connection can fail dependently as well, with a failure rate of $\lambda_{12dep} = 0.0022/y$ and an average repair time of $r_{12dep} = 10$ h. If both circuits fail dependently, both circuits are repaired simultaneously and put into operation together. If the circuits fail independently, they are repaired independently. Create the Markov model of this double circuit overhead line.

Answer Following the described approach to create a Markov model, the Markov model of this overhead line connection can be drawn as illustrated in Fig. 4.14. The repair rates are $\mu_{11} = \frac{1}{r_{11}/8760} = \frac{1}{8/8760} = 1095/y$ and $\mu_{12} = \frac{1}{r_{12}/8760} = \frac{1}{10/8760} = 876/y$ respectively.

From the normal operation state (S_1), circuit one can fail independently with failure rate λ_{11} (leading to state S_2), circuit two can fail independently with failure rate λ_{11} (leading to state S_3), or both circuits fail dependently with failure rate λ_{12dep} (leading to state S_4). From state S_2, circuit one can be repaired with repair rate μ_{11} (leading to state S_1), or circuit two can fail independently with failure rate λ_{11} (leading to state S_5). From state S_3, circuit two can be repaired with repair rate μ_{11} (leading to state S_1), or circuit one can fail independently with failure rate λ_{11} (leading to state S_5).

From state S_4, both circuits can be repaired simultaneously with repair rate μ_{12dep} (leading to state S_1). From state S_5, circuit two can be repaired with repair rate μ_{11} (leading to state S_2), or circuit one can be repaired with repair rate μ_{11} (leading to state S_3). The consequences for the transmission capacity of this double circuit (i.e. full capacity, half capacity, 0 capacity) are indicated by dashed lines in the Markov model.

In this example, when both circuits failed independently (state S_5), it is assumed that both circuits are repaired independently and at the same time. This means that the repair of one circuit will be completed first, after which the repair of the other circuit still needs to be completed. Consequently, the progress through the Markov model is either $S_5 \rightarrow S_2 \rightarrow S_1$ or $S_5 \rightarrow S_3 \rightarrow S_1$.

Fig. 4.14 Markov model of a double circuit overhead line (Example 4.6)

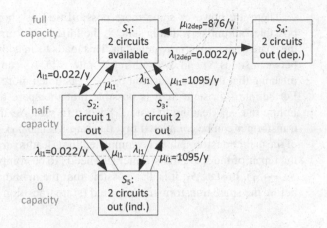

Here, the memorylessness of the exponential distribution is used (see also Example 3.3).

In reality, there are various other situations possible. For example, the simultaneous repair of both circuits, bringing the system back from state S_5 to state S_1 directly. Or the repair of the circuit by one repair team (one circuit after the other). These variations will lead to different Markov models, such that one must always be aware of the assumptions and conditions under which the Markov model holds.

Example 4.7: Markov Model Spare Transformer

Question Suppose that we have one transformer in a substation that supplies a load. This transformer has a failure rate of $\lambda_{t1} = 0.030/y$ and an average repair time of $r_{t1} = 168\,h$ (1 week). To reduce the impact of transformer failures, it is decided to install a spare transformer in the substation. This spare transformer can be connected manually to replace the failed transformer within an average switching time of $r_{sw} = 8\,h$. Create the Markov model of this spare transformer configuration.

Answer The Markov model of this situation can be developed in two steps. Figure 4.15 shows the first step. From the normal operation state (S_1), the transformer in operation can fail with failure rate λ_{t1} (leading to state S_2). From state S_2, the spare transformer can be connected with switching rate μ_{sw} (leading to state S_3). In state S_3, the failed transformer can be repaired with repair rate μ_{t1} (leading to state S_1). The consequence for the transmission capacity is indicated by the dashed gray line. This Markov model shows the normal progress of this spare transformer configuration.

There are however some more possibilities, as shown in Fig. 4.16. It is possible, but unlikely, that in state S_2 the failed transformer is repaired before the spare transformer is switched in. This leads to an additional state transition between states S_2 and S_1 (with repair rate μ_{t1}). As can be seen, $\mu_{sw} \gg \mu_{t1}$, meaning that the state transition $S_2 \rightarrow S_3$ is much more likely than $S_2 \rightarrow S_1$. Furthermore, in state S_3 it is possible that the spare transformer fails with failure rate λ_{t1} (leading to state S_4). Then in state S_4, the repair of the spare transformer can be completed first (leading to state S_3). Otherwise, the repair of the main transformer can be completed first (leading to state S_5), after which the repair of the spare transformer still needs to be completed (state transition $S_5 \rightarrow S_1$). In state S_5, it is also possible that the main transformer fails again before the spare transformer is repaired (state transition $S_5 \rightarrow S_4$).

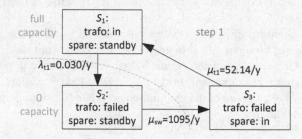

Fig. 4.15 Markov model of a transformer with a spare transformer (step 1) (Example 4.7)

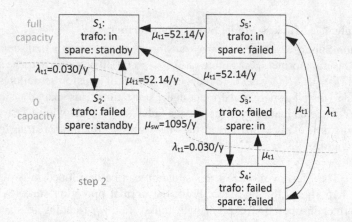

Fig. 4.16 Markov model of a transformer with a spare transformer (step 2) (Example 4.7)

4.2.5 Reduction Techniques

The examples in the previous section show that Markov models can quickly become complicated, even for small systems. For example, Fig. 4.17 shows the Markov model of two transformers and one spare transformer. Fortunately, there are several *reduction techniques* to simplify Markov models. With these techniques, it becomes possible to model somewhat larger systems with Markov models, while solving the Markov model becomes easier as well.

The following reduction techniques can be used:

1. When multiple state transitions lead to the same state from the same initial state, they can be combined to one transition with the sum of the individual transition rates (see Fig. 4.18) [3].
2. When there are multiple transitions that lead from the same initial state via intermediate states to the same final state, the intermediate states can be combined if (see Fig. 4.19) [3]:

   - all intermediate states belong to the same category;
   - all intermediate states lead with the same transition rate to the final state;
   - no other state transitions from/to the intermediate states exist.

3. When an event (with a certain transition rate) always leads to the same final state, irrespective of whether another event causes a temporary sidestep, the intermediate state of the sidestep can be removed if (see Fig. 4.20) [3]:

   - the initial state and the intermediate state belong to the same category;
   - no other state transitions from/to the intermediate state exist.

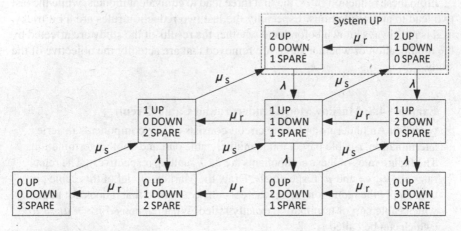

Fig. 4.17 More advanced state model

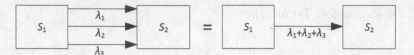

Fig. 4.18 Markov reduction technique 1

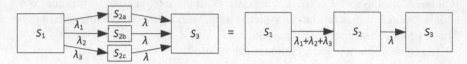

Fig. 4.19 Markov reduction technique 2

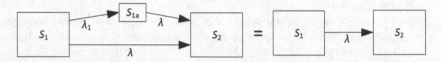

Fig. 4.20 Markov reduction technique 3

4. If there are multiple state transitions from the same initial state, the transitions
   with significantly smaller transition rates could be omitted from the model. This
   leads to a simplified approximation of the Markov model.
5. System states with very low probability could be omitted from the model if they
   are not the objective of the study. This leads to a simplified approximation of the
   Markov model.

From these reduction rules, the first three lead to equivalent models, while the last
two lead to approximations. Especially the last two reduction rules are a bit tricky,
as it is not always known beforehand whether the results of the study are affected by
the modification or whether states are removed that are actually the objective of the
study.

Example 4.8: Markov Model Underground Cable Circuit

Question An underground cable circuit consists of five components in series:
a termination, a cable part, a joint, another cable part and another termination.
The failure rates of these components are λ_t, λ_c and λ_j, respectively. The repair
rates are μ_t, μ_c and μ_j respectively. Draw the Markov model of this cable circuit. Apply the reduction rules, where it can be assumed that it does not matter
which cable part or termination exactly failed. What happens if $\mu_t = \mu_c = \mu_j$
(which can be called μ_{cs})?

Answer Because the components of the cable circuit are in series, the whole
circuit fails if one component fails. Then, the failed component is repaired

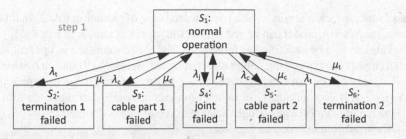

Fig. 4.21 Markov model of an underground cable circuit (Example 4.8)

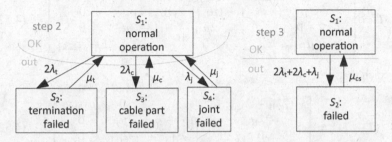

Fig. 4.22 Simplified Markov model of an underground cable circuit (Example 4.8)

and the whole circuit is brought into operation again. The Markov model then becomes as shown in Fig. 4.21. If it does not matter which cable part or termination exactly failed, the corresponding states can be combined. This leads to the Markov model as shown in Fig. 4.22 on the left.

If all components have the same repair rate (and repair time), and it does not matter which component exactly failed, all failure states can be combined. As shown in Fig. 4.22 on the right, this equals the two-state Markov model. This assumption makes further reliability analysis of networks with underground cable much easier.

Example 4.9: Reduced Markov Model Overhead Line

Question Apply reduction techniques to the Markov model of Example 4.6, where it can be assumed that if one circuit fails, it does not matter whether the failed circuit is circuit 1 or circuit 2. Solve the Markov model.

Answer When the Markov model of Example 4.6 is studied, it can be seen that states S_2 and S_3 are equivalent if it does not matter whether circuit 1 or circuit 2 has failed. In fact, it seems to be a complicated case of reduction rule 2. In this case, the Markov model can be reduced to the model as shown in Fig. 4.23.

This can also be explained as follows. When both circuits are in operation, both circuits can fail and the failure rate is doubled ($S_1 \rightarrow S_2$). If one circuit has failed, only one circuit can fail more ($S_2 \rightarrow S_3$) or one circuit can be repaired ($S_2 \rightarrow S_1$). If both circuits have failed, both can be repaired at the same time (if the repair team is large enough), such that repair rate is doubled ($S_3 \rightarrow S_2$).

The transition matrix then becomes:

$$T = \begin{bmatrix} 1 - (2\lambda_{11} + \lambda_{12\text{dep}}) & 2\lambda_{11} & 0 & \lambda_{12\text{dep}} \\ \mu_{11} & 1 - (\mu_{11} + \lambda_{11}) & \lambda_{11} & 0 \\ 0 & 2\mu_{11} & 1 - 2\mu_{11} & 0 \\ \mu_{12\text{dep}} & 0 & 0 & 1 - \mu_{12\text{dep}} \end{bmatrix} \qquad (4.52)$$

Such that:

$$(T - I)^{\text{tr}} = \begin{bmatrix} -(2\lambda_{11} + \lambda_{12\text{dep}}) & \mu_{11} & 0 & \mu_{12\text{dep}} \\ 2\lambda_{11} & -(\mu_{11} + \lambda_{11}) & 2\mu_{11} & 0 \\ 0 & \lambda_{11} & -2\mu_{11} & 0 \\ \lambda_{12\text{dep}} & 0 & 0 & -\mu_{12\text{dep}} \end{bmatrix} \qquad (4.53)$$

Replacing the lowest row and solving the next matrix equation:

$$\begin{bmatrix} -(2\lambda_{11} + \lambda_{12\text{dep}}) & \mu_{11} & 0 & \mu_{12\text{dep}} \\ 2\lambda_{11} & -(\mu_{11} + \lambda_{11}) & 2\mu_{11} & 0 \\ 0 & \lambda_{11} & -2\mu_{11} & 0 \\ 1 & 1 & 1 & 1 \end{bmatrix} \begin{bmatrix} P_{S_1} \\ P_{S_2} \\ P_{S_3} \\ P_{S_4} \end{bmatrix} = \begin{bmatrix} 0 \\ 0 \\ 0 \\ 1 \end{bmatrix} \qquad (4.54)$$

Gives:

$$P = \begin{bmatrix} 1.0000 \\ 4.02 \cdot 10^{-5} \\ 4.04 \cdot 10^{-10} \\ 2.51 \cdot 10^{-6} \end{bmatrix}^{\text{tr}} \qquad (4.55)$$

Consequently, one can expect that about 0.35 h/y only one circuit is available, and about 0.022 h/y both circuits are unavailable.

Example 4.10: Reduced Markov Model Spare Transformer

Question Describe how the reduction rules can be applied to the Markov model of Example 4.7. What are the consequences for the final results?

Fig. 4.23 Simplified
Markov model of a double
circuit overhead line
(Example 4.9)

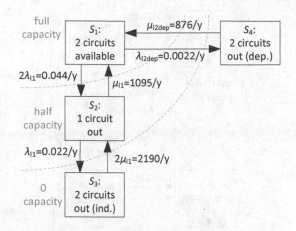

Answer In Example 4.7, the complete Markov model was shown in Fig. 4.16.
In the example, it was already mentioned that the state transition from state S_2
to state S_1 is much less likely than the state transition from state S_2 to state S_3.
Consequently, one could choose to omit this from the Markov model. This is
an example of reduction rule 4. Similarly, the transition from state S_5 to state
S_4 could be omitted.

One could also argue that it is not so likely that the spare transformer
fails during the short time that it is in use to replace the other transformer.
Consequently, states S_4 and S_5 could be removed then. This is an example
of reduction rule 5. This results in four possible Markov models, which are
depicted in Fig. 4.24.

To compare the differences between the results of the Markov model
approximations, the state probabilities of all four possible Markov model
approximations were calculated. The results are listed in Table 4.4. As can be
seen, there are small differences between the state probabilities of the model
approximations.

As we are finally interested in the probability that the load cannot be supplied, the probability of having no transmission capacity can be calculated by
adding the state probabilities of S_2 and S_4. The results are listed in Table 4.5.
As can be seen, there are only small differences between the four approximations. Omitting the state transition $S_2 \rightarrow S_1$ gives a somewhat larger error than
assuming that the spare transformer does not fail (omitting states S_4 and S_5).
In the end, all the approximations seem reasonable.

This example shows that it is a bit tricky to apply reductions rules 4 and 5 to
create approximations of the Markov model. Somehow, e.g. by experience or
verification, it must be known that these reductions do not cause large errors
in the final results. In this example, all four possibilities were analyzed to find
out that simplification of the Markov model is possible.

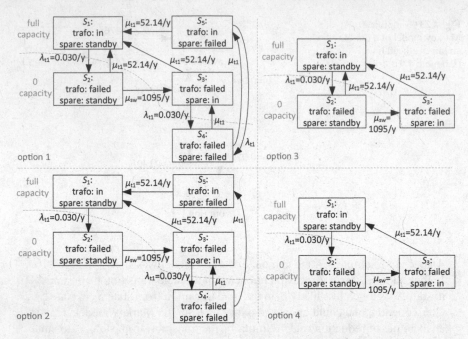

Fig. 4.24 Four possible Markov models of the transformer with spare transformer (Example 4.10)

Table 4.4 State probabilities of the four Markov models of Fig. 4.24 (Example 4.10)

	Option 1: with $S_2 \rightarrow S_1$ with S_4 & S_5	Option 2: without $S_2 \rightarrow S_1$ with S_4 & S_5	Option 3: with $S_2 \rightarrow S_1$ without S_4 & S_5	Option 4: without $S_2 \rightarrow S_1$ without S_4 & S_5
P_{S_1}	0.9994	0.99994	0.9994	0.9994
P_{S_2}	$2.61 \cdot 10^{-5}$	$2.74 \cdot 10^{-5}$	$2.61 \cdot 10^{-5}$	$2.74 \cdot 10^{-5}$
P_{S_3}	$5.49 \cdot 10^{-4}$	$5.75 \cdot 10^{-4}$	$5.49 \cdot 10^{-4}$	$5.75 \cdot 10^{-4}$
P_{S_4}	$1.58 \cdot 10^{-7}$	$1.65 \cdot 10^{-7}$	0	0
P_{S_5}	$1.58 \cdot 10^{-7}$	$1.65 \cdot 10^{-7}$	0	0

Table 4.5 Probability of having 0% capacity for the four Markov models of Fig. 4.24 (Example 4.10)

	Option 1: with $S_2 \rightarrow S_1$ with S_4 & S_5	Option 2: without $S_2 \rightarrow S_1$ with S_4 & S_5	Option 3: with $S_2 \rightarrow S_1$ without S_4 & S_5	Option 4: without $S_2 \rightarrow S_1$ without S_4 & S_5
$P[0\%]$	$2.63 \cdot 10^{-05}$	$2.76 \cdot 10^{-5}$	$2.61 \cdot 10^{-5}$	$2.74 \cdot 10^{-5}$

4.3 Fault Tree and Event Tree Analysis

System outages can be the result of various causes. For some systems, it can be complicated to study the impact of these various causes on the reliability by models like reliability networks and Markov models. In other cases, it might be desired to obtain knowledge about the direct relation between cause and effect. In these cases, an investigation of the component failures that finally lead to a system failure can be made by creating a *fault tree* or *event tree* [1, 4].

4.3.1 Fault Tree Analysis

Fault Tree Analysis (FTA) is an analytical technique which, by means of Boolean operators, describes how individual component failures can lead to a failure of the complete system. A fault tree consists of *events* and *Boolean operators*, as shown in Fig. 4.25.

In Fig. 4.26, two example fault trees are shown, one with an AND-gate and the other with an OR-gate. On top of the fault trees is the *top event*, which is a failure of the system in this case. On the bottom are the *basic events*, which are the component failures of component A and B in this case.

If the probabilities of the basic events are described by $P[A]$ and $P[B]$, the probability of the top event for the AND-gate can be calculated by (see also Appendix A) [5]:

$$P[S] = P[A \cap B] = P[AB] = P[A|B]P[B] \tag{4.56}$$

If events A and B are independent, $P[A|B] = P[A]$ and:

$$P[S] = P[A \cap B] = P[AB] = P[A]P[B] \tag{4.57}$$

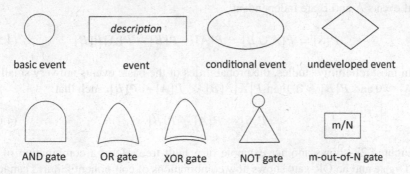

Fig. 4.25 Fault tree symbols

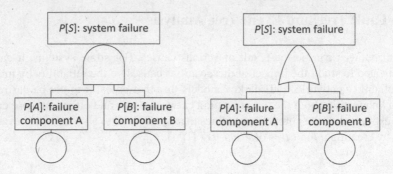

Fig. 4.26 OR-gate and AND-gate

Fig. 4.27 Example fault tree

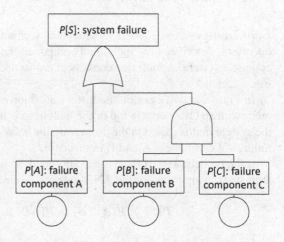

The probability of the top event for the OR-gate can be calculated by [5]:

$$P[S] = P[A \cup B] = P[A] + P[B] - P[AB] \tag{4.58}$$

If events A and B are independent:

$$P[S] = P[A \cup B] = P[A] + P[B] - P[A]P[B] \tag{4.59}$$

In most reliability studies, the probabilities of the basic events are very small. If $P[A] \approx 0$ and $P[B] \approx 0$, then $P[A]P[B] \ll P[A] + P[B]$, such that:

$$P[S] \approx P[A] + P[B] \tag{4.60}$$

Figure 4.27 shows another example of a fault tree. Here, a combination of an AND-gate and an OR-gate shows how combinations of component failures can lead to a failure of the system.

Assuming that the basic events are independent and assuming that the probabilities of the basic events are small, the probability of the top event can be calculated by:

$$P[S] \approx P[A] + P[B]P[C] \tag{4.61}$$

This is an approximation because of three reasons:

1. It is assumed that the basic events are independent.
2. It is assumed that the probabilities of the basic events are small.
3. The combinations of basic events are not mutually exclusive (e.g. the failure of components A, B and C together is counted twice: once as $P[A]$ and once as $P[B]P[C]$ as input of the OR-gate).

For the exact calculation of the top event, computer software could be used. In addition, solution methods like *Boolean algebra rules* and *minimal cut set* can be applied [1, 4].

Example 4.11: Fault Tree Offshore Network

Question Create a fault tree for the offshore network of Example 4.4 and calculate the probability that the network is unavailable. For convenience, the network configuration is shown again in Fig. 4.28, where the components of the network are labeled.

Answer Again, the fact that the offshore network is $N - 1$ redundant is crucial when studying the reliability of this particular network. By inspection of the network, there are basically four ways in which the network can fail (as also illustrated by the network cuts in Fig. 4.29)

1. (tr1a or cbl1a fails) and (tr1b or cbl1b fails)
2. (tr2a or cbl2a fails) and (tr2b or cbl2b fails)
3. (tr1a or cbl1a fails) and (br1 fails) and (tr2b or cbl2b fails)
4. (tr1b or cbl1b fails) and (br1 fails) and (tr2a or cbl2a fails)

The resulting fault tree then becomes as illustrated in Fig. 4.30. From the fault tree, the probability of a failure of the network can then be calculated as:

$$
\begin{aligned}
P[\text{network failure}] &\approx P[1] + P[2] + P[3] + P[4] \\
&\approx (U_{tr1} + U_{cbl})^2 + (U_{tr2} + U_{cbl})^2 + 2(U_{tr1} + U_{cbl})U_{br1}(U_{tr2} + U_{cbl}) \\
&\approx 2.5 \cdot 10^{-5} + 4.8 \cdot 10^{-6} + 1.1 \cdot 10^{-9} + 1.1 \cdot 10^{-9} \\
&\approx 3.0 \cdot 10^{-5} \tag{4.62}
\end{aligned}
$$

This is the same result as was calculated in Example 4.4. We can also see that the first term ($P[1]$) contributes the most to the final result. Therefore, the offshore transformers and cables (tr1 & cbl1) are the most critical components. As mentioned before, this result is an approximation. From the fault tree, it can be seen that some failure combinations are counted double. For example, the

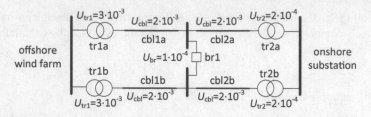

Fig. 4.28 Connection of an offshore wind farm by an offshore network (Example 4.11)

Fig. 4.29 Network cuts of
the offshore network of
Fig. 4.28 (Example 4.11)

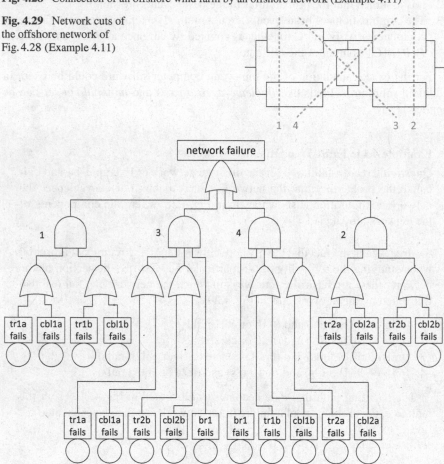

Fig. 4.30 Complete fault tree belonging to the offshore network of Fig. 4.28 (Example 4.11)

failure combination (tr1a, cbl1b, br1, tr2a) is part of AND-gate 1 and AND-gate 4. The probability of this failure combination is however too small to affect the final result.

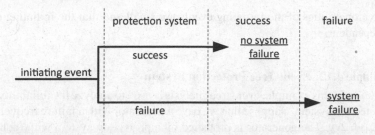

Fig. 4.31 Example event tree

4.3.2 Event Tree Analysis

Event tree analysis is a variation to fault tree analysis. The main difference is that fault tree analysis studies the combination of component failures that lead to a system failure, while event tree analysis studies the sequence of events that leads to a system failure [4, 6, 7]. It is therefore especially useful to study the response of a system to component failures, like the response of the protection system.

Figure 4.31 shows an example of an event tree. It can be seen that an event tree is drawn from left to right, in contrast to the fault tree, which is drawn from top to bottom. On the left of the event tree is the *initiating event*, which can be a component failure like a short-circuit. The initiating event will induce a *secondary event*, which can be the operation of the protection system. In this example, if the protection system works properly, there will not be a system failure. However, if the protection system fails, this will result in the *final event*: a failure of the system.

The probabilities of the final states of the event tree can be calculated from the probabilities of the initial event and the events that follow. Here, it must be realized that the probabilities of these secondary (and further) events are conditional probabilities (see also Appendix A). In the example of Fig. 4.31, the probability must be known that the protection system fails if the initial event occurs. The probability of a system failure can then be calculated by:

$$P[\text{system failure}] = P[\text{protection fails}|\text{initiating event}]P[\text{initiating event}] \quad (4.63)$$

Sometimes, the failure frequency of the initiating event is used instead of its probability. This is possible and results in the failure frequency of the final event as the other (conditional) probabilities are dimensionless:

$$f[\text{system failure}] = P[\text{protection fails}|\text{initiating event}]f[\text{initiating event}] \quad (4.64)$$

In most systems, there are many initiating events possible. Then, event trees are made for every possible initiating event. The results of these event trees are then combined to calculate the total probability of a system failure, for example by adding up the probabilities of a system failure. Similar to the fault tree, it is then assumed that the probabilities of the initiating events are small and that the initiating events are independent.

Example 4.12: Event Tree Protection System

Question In this example, event tree analysis is used to analyze the reliability of a protection system. Suppose that we have a generator with a failure frequency of $f_{g1} = 2/y$. The generator is protected by a protection system with (conditional) failure probability $P_{pr}[\text{failure}|\text{needed}] = 0.01$. As a backup, there is a backup protection, which has the same failure probability. If one of the protection systems successfully operates, only the generator is switched off. If both protection systems fail, a part of the substation to which the generator is connected is switched off. Draw the event tree of this situation. How often will only the generator be switched off? And how often will part of the substation be switched off?

Answer The event tree can be drawn as shown in Fig. 4.32. It clearly shows that if the generator fails and the primary protection works, the backup protection is not needed and only the generator is disconnected. If the primary protection fails and the backup protection works, only the generator is switched off. If both the primary and the backup protection fail, a part of the substation is switched off.

The failure frequencies of the final events are calculated by:

$$f_{g1out1} = f_{g1}(1 - P_{pr}[\text{failure}|\text{needed}]) = 2 \cdot 0.99 = 1.98/y \qquad (4.65)$$

$$f_{g1out2} = f_{g1}P_{pr}[\text{failure}|\text{needed}](1 - P_{pr}[\text{failure}|\text{needed}])$$
$$= 2 \cdot 0.01 \cdot 0.99 = 1.98 \cdot 10^{-2}/y \qquad (4.66)$$

$$f_{st1out1} = f_{g1}P_{pr}[\text{failure}|\text{needed}]P_{pr}[\text{failure}|\text{needed}]$$
$$= 2 \cdot 0.01 \cdot 0.01 = 2 \cdot 10^{-4}/y \qquad (4.67)$$

The frequency at which the generator is switched off is:

$$f_{g1out} = f_{g1out1} + f_{g1out2} = 1.98 + 1.98 \cdot 10^{-2} = 1.9998/y \qquad (4.68)$$

The frequency at which part of the substation is switched off is $f_{st1out1} = 2 \cdot 10^{-4}/y$

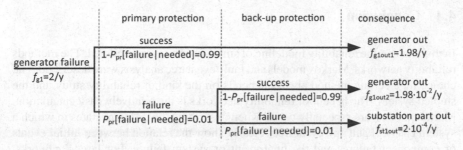

Fig. 4.32 Event tree of a protection system (Example 4.12)

Example 4.13: Event Tree Substation Configurations

In [6, 7], the reliability of various substation configurations was analyzed. The main objective of the study was to include the failure behavior of protection systems and compare the reliability of various possible substation configurations. The considered substation configurations were the typical double busbar configuration, the one-and-a-half circuit breaker configuration and the 4/3 circuit breaker configuration, as illustrated in Figs. 4.33, 4.34 and 4.35.

Normally, if a component fails, it is directly disconnected by the protection system. For example, when there is a short-circuit in section L_G1 of the 4/3 circuit breaker substation, circuit breakers CB_11 and CB_12 will operate. However, when circuit breaker CB_12 fails, CB_13 will open and line L1 will be disconnected as well. By investigating all possible failures of the protection system (i.e. circuit breaker failures as well as protection system failures), event trees were created. Figure 4.36 shows a typical event tree for this study.

The results from all event trees were collected and the probability of losing a certain number of connections (i.e. overhead lines, generators, loads) was compared for the three substation configurations. Table 4.6 and Fig. 4.37 show these results. It was concluded that although the probability of losing single lines/generators is the same for all substation configurations, there is a difference in the probability of losing multiple lines/generators. Because of the topology of the double busbar configuration, it is not possible to lose 2 lines/generators at the same time. It is however more likely to lose half or all of the lines/generators than in the other substation configurations. The differences between the substation configurations are especially large when considering the MTBF. It was concluded that the one-and-a-half and 4/3 circuit breaker configurations are preferred for substations that connect multiple generators.

4.4   Conclusion

In this chapter, the reliability modeling of small systems was discussed. The methods reliability networks, Markov models and fault/event tree analysis were described. The choice for a certain method always depends on the kind of reliability study and the studied system. The benefit of reliability networks is the relatively easy calculation. Markov models are especially useful when describing the different states in which a system can be. Fault and event trees clearly show the relation between initial events or component failures and the final event or system failure. For larger networks,

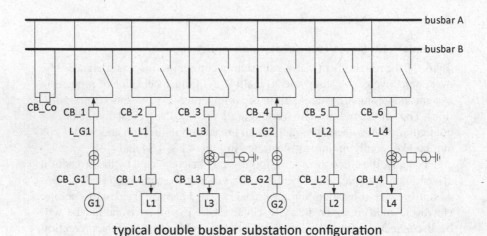

typical double busbar substation configuration

Fig. 4.33   Typical double busbar configuration (Example 4.13)

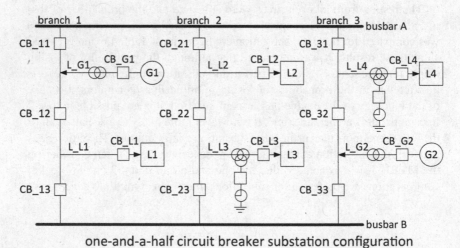

one-and-a-half circuit breaker substation configuration

Fig. 4.34   One-and-a-half circuit breaker configuration (Example 4.13)

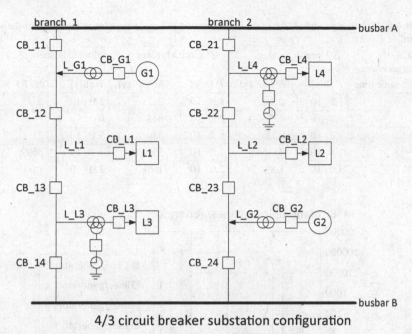

4/3 circuit breaker substation configuration

Fig. 4.35 4/3 circuit breaker configuration (Example 4.13)

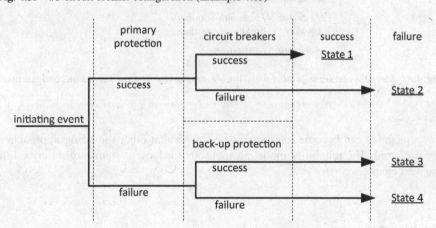

Fig. 4.36 Typical event tree of a substation protection system (Example 4.13)

Table 4.6 Probabilities and *MTBF* of the disconnection of lines/generators in the three substations (Example 4.13)

Lost lines/generators at the same time	3/2 circuit breaker		4/3 circuit breaker		double busbar	
	P [h/y]	*MTBF* [y]	P [h/y]	*MTBF* [y]	P [h/y]	*MTBF* [y]
1	$1.9 \cdot 10^2$	2	$1.9 \cdot 10^2$	2	$2.0 \cdot 10^2$	2
2	$8.8 \cdot 10^{-1}$	377	$2.2 \cdot 10^{-1}$	648	0	–
3	$7.2 \cdot 10^{-4}$	9206	$6.7 \cdot 10^{-1}$	840	$9.6 \cdot 10^{-1}$	21
6	$1.6 \cdot 10^{-4}$	41364	$1.1 \cdot 10^{-4}$	55000	$1.9 \cdot 10^{-3}$	3699
Total	$2.0 \cdot 10^{-2}$	1.81	$2.0 \cdot 10^2$	1.79	$2.0 \cdot 10^2$	1.63

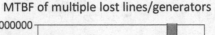

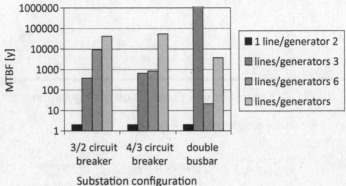

Fig. 4.37 Mean time between failures of lost lines/generators for the three substation configurations (Example 4.13)

these methods can become too complicated, such that other modeling approaches are preferred. The reliability modeling methods for large systems are described in the next chapter.

Problems

4.1 Learning Objectives
Please check the learning objectives as given at the beginning of this chapter.

4.2 Main Concepts
Please check whether you understand the following concepts:

- Reliability networks:
  - Series network / parallel network;

- 1-out-of-N / m-out-of-N / N-out-of-N;
- Redundancy and dependent failures;
- Decomposition method.

- Markov models:

  - System state (S);
  - (State) transition rate and frequency;
  - Transition matrix and state probability vector;
  - Equilibrium;
  - Reduction techniques.

- Fault/event tree analysis:

  - Fault tree (analysis);
  - Events / boolean operators;
  - Top event / basic event;
  - Event tree (analysis);
  - Initial event / secondary event / final event.

4.3 Unavailability Cable Circuit

A 220 kV underground cable circuit consists of a single (three-core) cable. The cable circuit is 20 km long and consists of cable parts of 1000 m. Terminations are installed at the cable circuit ends, while joints are installed between the cable parts. Draw the layout of this cable circuit. The failure rates of the cable parts, joints and terminations are $\lambda_c = 0.003/$cctkmyr, $\lambda_j = 0.002/$compyr and $\lambda_t = 0.001/$compyr, and the repair time of all components is $r_{cs} = 336$h. What are the failure frequency and the unavailability of this cable circuit? Two of these cable circuits now form a double circuit. What are the probabilities that both circuits are available, that only one circuit is available and that both circuits are unavailable?

4.4 Reliability Offshore Network

In Example 4.4, it was mentioned that the fact that this particular offshore network is $N - 1$ redundant is crucial when solving the network. Verify yourself what exactly will happen if this offshore network is not $N - 1$ redundant.

4.5 Reliability Network Cable

How is Example 4.8 related to reliability networks?

4.6 Markov Model Generators

Suppose we have a power plant with two generators. One generator is normally used, while the second generator is a spare. The failure rates of both generators are $\lambda_g = 4/$y, while the repair time is $r_g = 24$h. The switching time to the spare generator is $r_{sw} = 4$h. Draw a Markov model of this situation, where you can assume that the first generator is not repaired during the switching time and that it is very unlikely that the spare generator fails. What exactly will change in the Markov model if these two assumptions are not made?

4.7 Markov Model Variations

In Example 4.6, it was mentioned that a different description and modeling of reality leads to a different Markov model. Develop the Markov model of Example 4.6 when:

- if both circuits fail independently, they are repaired simultaneously (meaning that the repair of both circuits is completed at the same time) with repair rate μ_{12ind};
- if both circuits fail independently, they are repaired by one repair team (one after the other). You can assume that the repair team first completes the repair of the circuit that failed first.

4.8 Markov Model Solution

Solve the Markov models as shown in Example 4.10.

4.9 Fault/Event Tree Analysis

Describe fault tree analysis and event tree analysis. What are the similarities and differences? What assumptions are often made when calculating the final results?

4.10 Fault Tree Analysis

Solve Example 4.2 with Fault Tree Analysis.

4.11 Event Tree Analysis

Suppose a generator is connected to a substation and protected by a protection system. If there is a short-circuit in the generator, the generator is switched off and disconnected by the protection system. The protection system can fail in two ways: the circuit breaker of the generator does not open or the fault is not detected. If the circuit breaker does not open, another circuit breaker will operate and the generator is successfully disconnected. If the fault is not detected, the protection system of the substation will switch off half the substation. Draw an event tree of this situation. The failure frequency of (a short-circuit in) the generator is $f_g = 5/y$, the conditional failure probability of the circuit breaker is $P_{cb} = 0.05$, and the conditional failure probability that the fault is not detected is $P_{ps} = 0.01$. How frequent will the second circuit breaker be needed? And how frequent will half the substation be switched off?

References

1. Klaassen KB, van Peppen JCL, Bossche A (1988) Bedrijfszekerheid, Theorie en Techniek. Delftse Uitgevers Maatschappij, Delft, the Netherlands
2. Li W (2005) Risk assessment of power systems-models, methods, and applications. Wiley Interscience - IEEE Press, Canada
3. Bossche A (2016) Microelectronics Reliability - Lecture Notes. Delft University of Technology, Delft, the Netherlands
4. Cepin M (2011) Assessment of power system reliability - methods and applications. Springer, London
5. Yates RD, Goodman DJ (2005) Probability and stochastic processes. Wiley, United States

6. Wang F, Tuinema BW, Gibescu M, Meijden van der MAMM (2012) Reliability evaluation of substations subject to protection system failures. MSc Thesis, Department of Electrical Engineering, Delft University of Technology, Delft, the Netherlands
7. Wang F, Tuinema BW, Gibescu M, Meijden van der MAMM (2013) Reliability evaluation of substations subject to protection system failures. Powertech2013, Grenoble, France

Chapter 5
Reliability Models of Large Systems

In the previous chapters, reliability models of individual components and small systems were discussed. It was already mentioned that it easily becomes too complicated to combine these models to create a reliability model of a large system. Therefore, for large systems special reliability analysis approaches have been developed. This chapter discusses three of these approaches. The first, state enumeration, is an analytical analysis of possible system failure states. State enumeration will be discussed in detail in Sect. 5.1. The second approach, generation adequacy, focuses on the reliability of generation systems and is discussed in Sect. 5.2. The third, Monte Carlo simulation, is a (computer) simulation approach of the system behavior. This will be discussed in Sect. 5.3.

Learning Objectives

After reading this chapter, the student should be able to:
- Describe state enumeration and apply it to study the reliability of large systems;
- Describe and perform generation adequacy analysis for generation systems;
- Describe Monte Carlo simulation and apply it to analyze the reliability of large systems.

© Springer Nature Switzerland AG 2020

B. W. Tuinema et al., *Probabilistic Reliability Analysis of Power Systems*,
https://doi.org/10.1007/978-3-030-43498-4_5

5.1 State Enumeration

In Sect. 4.2, the application of Markov models to describe the possible system states was ·discussed. For larger systems, the combination of the possible states of all components would lead to an extremely complicated model with a huge amount of states. Therefore, special reliability methods exist for larger systems. One of these approaches is state enumeration. In this method, a selection is made of the most likely system states. This selection is assumed to be representative for the overall system reliability. The following example shows how the number of possible system states increases significantly for larger power systems.

Example 5.1: Number of System States

Question The number of system states increases with the size of the system. This can be explained by a simple calculation. Suppose that a system consists of 100 connections (about the size of the Dutch EHV transmission network). If every connection in the system can be in or out of service, each individual connection could be modeled by a two-state Markov model (see also Sect. 3.3). For simplicity, other components of the transmission network are not considered and dependent failures are omitted. If each connection has two states, what is the number of system states for the complete Dutch EHV transmission network? How many possibilities of 0, 1, 2, 3, 4 or 5 connection failures exist?

Answer The total number of system states can be calculated by:

$$n^N = 2^{100} = 1.3 \cdot 10^{30} \text{ states} \tag{5.1}$$

where
N = number of components in the system [–]
n = number of states per component [–]
  The number of possible combinations for a certain number of failures in the system can be calculated by the mathematical combinations:

Table 5.1 Number of system states per failure order (Example 5.1)

Failure order	System states
0th	1
1st	100
2nd	4950
3rd	$1.6 \cdot 10^5$
4th	$3.9 \cdot 10^6$
5th	$7.5 \cdot 10^7$

$$\binom{N}{k} = \frac{N!}{k!(N-k)!} \tag{5.2}$$

where

N = number of components in the system [–]
k = number of component failures [–]

$$\binom{100}{1} = 100 \tag{5.3}$$

$$\binom{100}{2} = \frac{100 \cdot 99}{2 \cdot 1} = 4950 \tag{5.4}$$

$$\binom{100}{3} = \frac{100 \cdot 99 \cdot 98}{3 \cdot 2 \cdot 1} = 1.6 \cdot 10^5 \tag{5.5}$$

Table 5.1 gives an overview of the number of possible system states per failure order. It is clear that the number of possible states increases considerably for higher-order failure states. In addition to this large number of possible failure states, one has to be aware of the fact that load/generation scenarios consist of multiple states as well. If one studies the reliability of a power system using an hourly year scenario for the generation/load, one has to analyze the possible failure states for all 8760 h of the year.

5.1.1 Deterministic Contingency Analysis

As the approach of state enumeration is quite similar to deterministic contingency analysis, it is helpful to discuss the latter in detail first. In *deterministic contingency analysis*, components are taken out of the power system in a simulation. It is analyzed whether component failures lead to serious issues in the power system, like component overloading or over- and undervoltages. The followed procedure is illustrated in Fig. 5.1. Often, the results of the simulations are used to determine whether the system meets certain *deterministic criteria*. For example, the $N-1$ *redundancy criterion* states that a failure of a single component must not result in a failure of the system.

Deterministic contingency analysis is performed up to a certain level. Usually, single contingencies and certain critical (dependent) double contingencies are considered. The number of higher-order contingencies is too large to analyze, while the probabilities of these failure states are usually assumed to be very small.

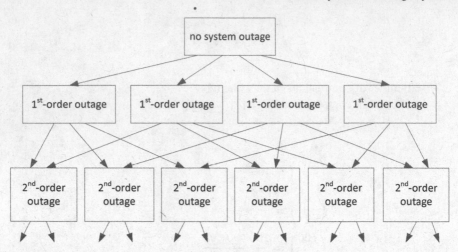

Fig. 5.1 Approach of contingency analysis and state enumeration

5.1.2 Probabilistic Reliability Indicators

State enumeration follows an approach quite similar to deterministic contingency analysis. In state enumeration, system states are defined according to their order of failure as well, as illustrated in Fig. 5.1. The lower-order states, with less component failures and a higher probability, are considered first. Then, the higher-order states, with more failures and lower probability, are considered. It is assumed that the lowest-order states, with the highest probability, are representative for the system reliability. State enumeration is mainly used in systems with low failure probabilities and when no additional information like probability distributions or switching sequences of protection systems is required.

The main difference with deterministic contingency analysis is that in state enumeration, the state probabilities and the effects of the states are combined to calculate probabilistic reliability indicators. As already mentioned in Sect. 1.2, the security of supply probably is the most important reliability indicator. It is therefore of interest to determine how often, how long, and how much load is disconnected. Some probabilistic reliability indicators directly related to this are [1–3]:

- *PLC*: *Probability of Load Curtailment* [–]
  The probability that the demanded load cannot be supplied (partially or completely). The PLC is often determined based on a set of considered contingencies (single contingencies as well as higher-order contingencies) and a studied time period (usually on a yearly basis).

$$PLC = \sum_{i \in S_c} (P[\text{load curtailment}|i]P_i) = \sum_{i \in S_c} \left(\frac{t_i}{T} P_i \right) \tag{5.6}$$

where
S_c = set of considered contingencies
$P[\text{load curtailment}|i]$ = probability that contingency i causes load curtailment [–]
P_i = probability of contingency i [–]
t_i = total time that there is load curtailment during the studied period, given that contingency i is present [h]
T = total time of the studied period [h]

- **EENS**: *Expected Energy Not Supplied* [MWh]
  Total amount of energy that is expected not to be supplied during a given time period (usually on a yearly basis) due to supply interruptions.

$$EENS = \sum_{i \in S_c} (E_i P_i) \tag{5.7}$$

where
S_c = set of considered contingencies
E_i = total curtailed energy in the study period if contingency i is present [MWh]
P_i = probability of contingency i [–]

- **SAIDI**: *System Average Interruption Duration Index* [min]
  Average outage duration of each customer during a given time period (usually on a yearly basis).

$$SAIDI = \frac{\sum_i^n (r_i N_i)}{N_t} \tag{5.8}$$

where
n = number of interruptions [–]
r_i = duration of interruption i [min]
N_i = number of customers not served due to interruption i [–]
N_t = total number of customers [–]

- **SAIFI**: *System Average Interruption Frequency Index* [–]
  Average number of interruptions per customer during a given time period (usually on a yearly basis).

$$SAIFI = \frac{\sum_i^n N_i}{N_t} \tag{5.9}$$

where
n = number of interruptions [–]
N_i = number of customers not served due to interruption i [–]
N_t = total number of customers [–]

- **CAIDI**: *Customer Average Interruption Duration Index* [min]
  Average interruption duration per interrupted customer during a given time period (usually on a yearly basis).

$$CAIDI = \frac{\sum_i^n (r_i N_i)}{\sum_i^n N_i} = \frac{SAIDI}{SAIFI} \tag{5.10}$$

where

n = number of interruptions [–]

r_i = duration of interruption i [min]

N_i = number of customers not served due to interruption i [–]

As already discussed in Chap. 1, the interruption of consumers is not the only risk for a TSO. The disconnection of producers and generator redispatch can be a (financial) risk as well. Therefore, additional probabilistic reliability indicators can be defined [1]:

- *Probability of Overload* [–]
  The probability that one or more connections in the network are overloaded during a given time period (usually on a yearly basis).

$$P_{\text{overload}} = \sum_{i \in S_c} (P[\text{overload}|i]P_i) = \sum_{i \in S_c} \left(\frac{t_{\text{overload},i}}{T} P_i \right) \tag{5.11}$$

where

S_c = set of considered contingencies

$P[\text{overload}|i]$ = probability that contingency i causes an overloaded connection [–]

P_i = probability of contingency i [–]

$t_{\text{overload},i}$ = total time that there is an overload during the studied period, given that contingency i is present [h]

T = total time of the studied period [h]

- *Probability of Generation Redispatch* [–]
  The probability that generation redispatch is applied during a given time period (usually on a yearly basis).

$$P_{\text{redispatch}} = \sum_{i \in S_c} (P[\text{redispatch}|i]P_i) = \sum_{i \in S_c} \left(\frac{t_{\text{redispatch},i}}{T} P_i \right) \tag{5.12}$$

where

S_c = set of considered contingencies

$P[\text{redispatch}|i]$ = probability that contingency i causes generation redispatch [–]

P_i = probability of contingency i [–]

$t_{\text{redispatch},i}$ = total time that redispatch is applied during the studied period, given that contingency i is present [h]

T = total time of the studied period [h]

- *Expected Redispatch Costs* [currency]
  Expected costs of generation redispatch during a time period (usually on a yearly basis).

$$C_{\text{redispatch}} = \sum_{i \in S_c} \left(C_{\text{redispatch},i} P_i \right) \qquad (5.13)$$

where
S_c = set of considered contingencies
$C_{\text{redispatch},i}$ = total redispatch costs during the study period,
· given that contingency i is present [currency]
P_i = probability of contingency i [–]

- *Probability of the Alert State* [–]
The probability that the power system is in the alert state (i.e. $N - 0$ state) during a given time period (usually on a yearly basis).

$$P_{\text{alert}} = \sum_{i \in S_c} (P[\text{alert}|i]P_i) = \sum_{i \in S_c} \left(\frac{t_{\text{alert},i}}{T} P_i \right) \qquad (5.14)$$

where
S_c = set of considered contingencies
$P[\text{alert}|i]$ = probability that contingency i leads to the alert state [–]
P_i = probability of contingency i [–]
$t_{\text{alert},i}$ = total time that the network is in the alert state during the studied period, given that contingency i is present [h]
T = total time of the studied period [h]

Example 5.2: Simple State Enumeration

Question Assume we have a transmission network consisting of only four overhead lines, with unavailabilities as listed in Table 5.2. A state enumeration is performed up to the second-order failures. Only independent failures are considered. The objective is to find the probability and amount of load curtailment because of overhead line failures. It is found that single overhead line failures do not lead to load curtailment. Furthermore, it is found that a failure of OHL1 and OHL3 together leads to 100 MW load curtailment, while a failure of OHL2 and OHL4 results in 500 MW load curtailment. This situation holds for 1000 h/y, while the rest of the year there are no single and independent double contingencies that lead to load curtailment.

Sketch how the state enumeration looks like. Is this transmission network $N - 1$ redundant? Is it $N - 2$ redundant? If not, how many h/y is the network not $N - 2$ redundant? What is the Probability of Load Curtailment (*PLC*)? What is the Expected Energy Not Supplied (*EENS*)? What is the total probability of the failure states that are not considered in this state enumeration?

Answer The state enumeration can be sketched as shown in Fig. 5.2. In the sketch, it is indicated which states cause load curtailment. As no single component failures cause load curtailment (and load curtailment is the objective of this particular study), this transmission network can be regarded as $N-1$ redundant. As there are independent double contingencies that lead to load curtailment, this network is not $N-2$ redundant. This situation holds for 1000 h/y, while the rest of the year there are no single and independent double contingencies that lead to load curtailment, so this network is 1000 h/y not $N-2$ redundant.

The Probability of Load Curtailment can be calculated by:

$$
\begin{aligned}
PLC &= \sum_{i \in S_c} \left(\frac{t_i}{T} P_i \right) = \frac{1000}{8760} U_1 A_2 U_3 A_4 + \frac{1000}{8760} A_1 U_2 A_3 U_4 \\
&= \frac{1000}{8760} U_1 (1 - U_2) U_3 (1 - U_4) + \frac{1000}{8760} (1 - U_1) U_2 (1 - U_3) U_4 \\
&= \frac{1000}{8760} \cdot 2 \cdot 10^{-5} \cdot 0.99996 \cdot 6 \cdot 10^{-5} \cdot 0.99992 \\
&\quad + \frac{1000}{8760} \cdot 0.99998 \cdot 4 \cdot 10^{-5} \cdot 0.99994 \cdot 8 \cdot 10^{-5} = 5.0 \cdot 10^{-10}
\end{aligned}
$$

$$(5.15)$$

This equals $4.4 \cdot 10^{-6}$ h/y.

If we define the duration of the load curtailment ($r_{curt} = 1000$ h/y) and the amount of curtailed load ($L_{curt,1,3} = 100$ MW and $L_{curt,2,4} = 500$ MW), the Expected Energy Not Supplied becomes:

$$
\begin{aligned}
EENS &= \sum_{i \in S_c} (E_i P_i) = r_{curt} L_{curt,1,3} U_1 A_2 U_3 A_4 + r_{curt} L_{curt,2,4} A_1 U_2 A_3 U_4 \\
&= 1000 \cdot 100 \cdot 2 \cdot 10^{-5} \cdot 0.99996 \cdot 6 \cdot 10^{-5} \cdot 0.99992 \\
&\quad + 1000 \cdot 500 \cdot 0.99998 \cdot 4 \cdot 10^{-5} \cdot 0.99994 \cdot 8 \cdot 10^{-5} = 1.7 \cdot 10^{-3}
\end{aligned}
$$

$$(5.16)$$

The $EENS$ of $1.7 \cdot 10^{-3}$ MWh/y is very small.

In the equations above, one might choose to omit the component availabilities as $A_n \approx 1$. This is indeed possible for small systems and for very small unavailabilities, but one must be aware that for larger systems the product of the availabilities mostly is not approximately 1 anymore (e.g. $0.999^{100} = 0.9$). Therefore, it is recommended to always include the component availabilities when calculating the state probabilities in state enumeration. This also enables us to check whether the state enumeration is correct, as the sum of all the state probabilities must be 1.

The total probability of the not-considered states is:

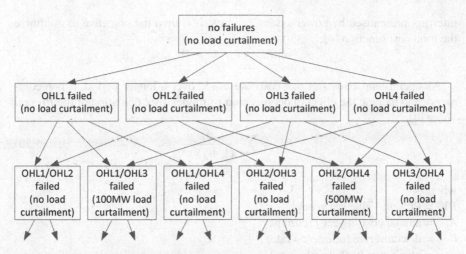

Fig. 5.2 State enumeration of the network consisting of four transmission lines (Example 5.2)

Table 5.2 Unavailabilities of the four overhead lines (Example 5.2)

Overhead line no.	OHL1	OHL2	OHL3	OHL4
Unavailability (U_n) [–]	$2 \cdot 10^{-5}$	$4 \cdot 10^{-5}$	$6 \cdot 10^{-5}$	$8 \cdot 10^{-5}$

$$P_{\text{error}} = U_1 U_2 U_3 U_4 + A_1 U_2 U_3 U_4 + U_1 A_2 U_3 U_4 + U_1 U_2 A_3 U_4$$
$$+ U_1 U_2 U_3 A_4 = 4.0 \cdot 10^{-13} \qquad (5.17)$$

This equals 1 minus the total probability of the considered states:

$$P_{\text{error}} = 1 - A_1 A_2 A_3 A_4 - U_1 A_2 A_3 A_4 - A_1 U_2 A_3 A_4$$
$$- A_1 A_2 U_3 A_4 - A_1 A_2 A_3 U_4 - U_1 U_2 A_3 A_4 - U_1 A_2 U_3 A_4$$
$$- U_1 A_2 A_3 U_4 - A_1 U_2 U_3 A_4 - A_1 U_2 A_3 U_4 - A_1 A_2 U_3 U_4 \qquad (5.18)$$

5.1.3 Probabilistic Cost Analysis

In economical studies, it is of interest to compare the total costs and benefits of an investment. Typically, the *investment costs (I)*, the *operational costs (O)*, and the *risk costs (R)* are determined and compared [4]. The investment costs are the CAPEX (capital expenditure) as mentioned before in Sect. 1.4, while the operational costs are equivalent to the operational expenditure (OPEX). The risk costs are the costs of interruptions caused by power system failures. It is then the objective to minimize the total cost function [4]:

$$\min T = I + O + R \tag{5.19}$$

As operational costs and risk costs are made over a longer period, it is recommended to include the effect of depreciation of money by calculation of the *Present Value* [4]:

$$PV = \sum_{j=1}^{m} \frac{A_j}{(1+i)^{j-1}} \tag{5.20}$$

where
PV = Present Value [currency]
A_j = annual cost in year j [currency]
i = discount rate (or interest rate) [–]
m = number of considered years [–]

In this way, all costs are projected to the first year, such that a comparison of the costs can be made. Equation (5.19) becomes (assuming that investment costs are only made in year 1):

$$\min T = I + O_{PV} + R_{PV} \tag{5.21}$$

Minimizing the total cost function will result in the economically most optimal solution. In many practical cases however, the objective is to determine how much the current situation improves if a certain development is implemented. The new development (e.g. the installation of a new overhead line connection) has certain investment costs and operational costs, but will lead to a reduction of the risk costs. It is then of interest to calculate whether the investment and operational costs can be earned back by the reduced risk costs. By calculating the *Net Present Value*, it can be determined whether the development is economical:

$$NPV = R_{PV,\text{impr}} - O_{PV} - I \tag{5.22}$$

where
NPV = Net Present Value [currency]
$R_{PV,\text{impr}}$= improvement of the (Present Value) risk costs [currency]

The profitability of a certain investment can then be indicated by a (discounted) *Return on Investment (ROI)*:

$$ROI = \frac{NPV}{I} \tag{5.23}$$

5.1.4 State Enumeration of Large Power Systems

State enumeration of large power systems can effectively be performed by computer programs. These programs normally include the calculation of failure state probabilities, load flow calculations, the application of remedial actions, and the presentation of the results. Figure 5.3 shows a typical state enumeration algorithm.

The algorithm analyzes each contingency that has to be considered. Contingencies are often described in contingency lists, e.g. a list of independent single contingencies and a list of dependent double contingencies. Failure states up to a certain order are analyzed, such that, for example, independent single contingencies, dependent double contingencies, and combinations of two of these are considered.

For each studied contingency, a certain snapshot of the system generation and load is considered. This can, for example, be a specific hour of a year scenario, but could also be sampled from a set of generation/load snapshots. It is recommended to follow the order (for every contingency)-(for every hour), rather than (for every hour)-(for every contingency). Because for each studied contingency the grid topology changes, the grid model needs to be modified which takes much computation time. For different generation/load snapshots, the grid model does not change, such that it is faster to analyze every hour per studied contingency.

Then, a load flow is performed. This can be a DC load flow, if only active power is to be considered, or an AC load flow, if reactive power and voltage behavior is to be studied as well. If any problems in the network are found, like component overloading or over-/undervoltages, remedial actions must be applied to relieve the network. Typical remedial actions are local generation redispatch, international generation redispatch, and load curtailment. Remedial actions are performed until the load flow analysis does not show problems in the network anymore.

The state enumeration proceeds until all generation/load snapshots and all-to-be-studied contingencies are considered. The results of the remedial actions are then collected, and reliability indicators are calculated.

Example 5.3: State Enumeration Program

Question Design a computer program algorithm that performs state enumeration. The program must:

- Study independent single contingencies, dependent double contingencies, and combinations of two independent single contingencies;
- Perform a DC load flow for every hour of the year and test whether the network is overloaded;
- Apply generation redispatch if needed, other remedial actions are not needed;
- Calculate the probability and expected amount of generation redispatch;
- Calculate the error, in the sense of total probability of not-considered states.

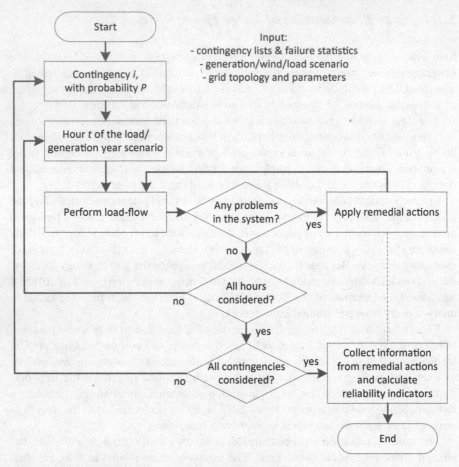

Fig. 5.3 Typical algorithm of state enumeration

Answer The precise implementation depends on the software that is used, but the general algorithm will be:

% *initial values*
$P_{\text{error}} = 1$ (this will be the total probability of not-considered states)
$P_{\text{redispatch}} = 0$ (this will be the probability of redispatch)
$E_{\text{redispatch}} = 0$ (this will be the expected amount of redispatch)

% *0th-order: normal system operation*
$P_{\text{state}} = $ calculate state probability
$P_{\text{error}} = P_{\text{error}} - P_{\text{state}}$
create the network model

```
for(every hour of the year)
    perform dc load flow
    if(network is overloaded)
        P_redispatch = P_redispatch + P_state/8760
        perform generation redispatch (E_rd = amount for 1 hour)
        E_redispatch = E_redispatch + E_rd · P_state
    end if
end for
```

% *1st-order: independent single contingencies*
```
for(every independent single contingency)
    P_state = calculate state probability
    P_error = P_error - P_state
    create the network model
    for(every hour of the year)
        perform dc load flow
        if(network is overloaded)
            P_redispatch = P_redispatch + P_state/8760
            perform generation redispatch (E_rd = amount for 1 hour)
            E_redispatch = E_redispatch + E_rd · P_state
        end if
    end for
end for
```

% *2nd-order: dependent double contingencies*
```
for(every dependent double contingency)
    P_state = calculate state probability
    P_error = P_error - P_state
    create the network model
    for(every hour of the year)
        perform dc load flow
        if(network is overloaded)
            P_redispatch = P_redispatch + P_state/8760
            perform generation redispatch (E_rd = amount for 1 hour)
            E_redispatch = E_redispatch + E_rd · P_state
        end if
    end for
end for
```

% *2nd-order: 2 independent single contingencies*
```
for(every independent single contingency)
    for(every (other) independent single contingency)
        P_state = calculate state probability
        P_error = P_error - P_state
```

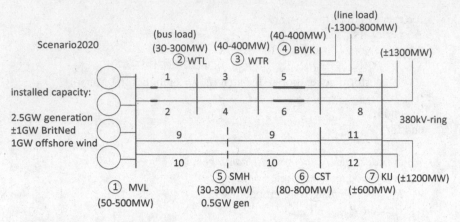

Fig. 5.4 Network configuration around Maasvlakte (MVL) substation (Example 5.4)

```
create the network model
for(every hour of the year)
    perform dc load flow
    if(network is overloaded)
        P_redispatch = P_redispatch + P_state/8760
        perform generation redispatch (E_rd = amount for 1 hour)
        E_redispatch = E_redispatch + E_rd · P_state
    end if
end for
end for
```

% results of state enumeration
$P_{redispatch}$ = probability redispatch in [−]
$8760 \cdot P_{redispatch}$ = probability redispatch in [h/y]
$E_{redispatch}$ = expected amount of redispatch in [MWh/y]
P_{error} = total probability of not-considered states

It must be mentioned here that if dependent double contingencies are included in a state enumeration, this might lead to a small error in the results. The main issue is that in reality, if for example overhead lines 1 and 2 fail dependently, overhead line 2 cannot fail independently anymore because it is switched off. In state enumeration, this situation (OHL1&2 failed dependently and OHL2 fails independently) might be included. In practice, the error caused by this is mostly small and can be neglected.

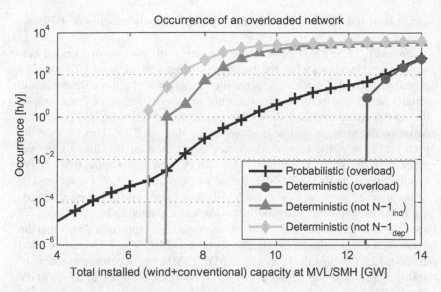

Fig. 5.5 Probability of an overloaded network or the $N - 0$ state in the Maasvlakte region network and comparison with deterministic analysis (Example 5.4)

Example 5.4: State Enumeration Maasvlakte Region

In [5], the reliability of transmission networks with EHV underground cables and offshore networks for wind energy was studied. Part of the research was a comparison of deterministic and probabilistic reliability indicators. For this study, a part of the Dutch 380kV transmission network was considered, namely the region around Maasvlakte (MVL380) substation. This part of the EHV transmission network is illustrated in Fig. 5.4. In the network diagram, the capacity of the generation and the ranges of substation/line loads are indicated. As can be seen, in the base scenario there is 1 GW installed offshore wind capacity connected to MVL380 substation. The location of the Randstad380 Zuid underground cable is indicated by bold lines.

A state enumeration was performed using a generation/wind/load scenario for the year 2020. To study the reliability impact of increasing offshore wind capacity, the installed capacity of the offshore wind was increased while it was assumed that the extra wind power is absorbed in the 380kV-ring connected to KIJ380. Figure 5.5 shows the main results of this study. The graph is an illustrative example of the difference between deterministic and probabilistic approaches. It shows the probability of an overloaded network (black line, +) and the deterministic occurrence of an overloaded network (dark gray, ●). Furthermore, the graph shows how often the network is not deterministic $N - 1$ redundant, if only independent contingencies are considered (light gray, ▲), and if both independent and dependent double circuit are considered (light gray, ◆).

From Fig. 5.5, it can be seen that the probability of an overloaded network indeed increases when the loading of the network increases. To compare probabilistic and deterministic reliability indicators, the three deterministic criteria and the probabilistic indicator are shown in one graph. For example, the deterministic overload shows that at about 12.5 GW installed generation capacity, the network will be overloaded for sure for 2 h/y. This because the total load flow in the networks becomes larger than the maximum transport capacity of the network. Furthermore, it can be seen that at about 6.5–7.0 GW installed capacity at MVL/SMH, the network is 1 h/y not $N - 1$ redundant. Consequently, the network is in the $N - 0$ state (alert state). However, according to the probabilistic analysis, the network is expected to be overloaded for about $1 \cdot 10^{-3}$ h/y in this case. Also, the deterministic approach shows that the network is 1 h/y in the $N - 0$ state at about 6.5–7.0 GW and 1 h/y overloaded at about 12.5 GW installed capacity at MVL/SMH, but according to the probabilistic approach, the network will be overloaded for 1 h/y at about 9 GW installed capacity.

The difference between deterministic and probabilistic reliability indicators is illustrated by this example. At the same time, it shows the challenges of interpreting the results of reliability analysis. If it is decided to use a probabilistic criterion, what probability of an overloaded network do we accept and how much generation capacity can then be installed at MVL380 substation? Of course, this also depends on the remedial actions that are needed to relieve the network overload. Will generation redispatch be sufficient or will load be curtailed?

5.1.5   State Enumeration as a (Partial) Markov Model

In the previous section, it was explained how to perform a state enumeration, while in Sect. 4.2, it was explained how to create Markov models for small systems. In both cases, an overview of the possible system states is made. In this sense, it can be asked whether state enumeration can be seen as a (partial) Markov model. In fact, state enumeration shares some characteristics with Markov models. System states and state transitions are described in both methods. In both methods, the probabilities of the system states can be calculated as the product of the component (un)availabilities. Also, the equation for the state transition frequencies (4.30) can be applied to state enumeration:

$$f_{S_i \to S_j} = P_{S_i} h_{ij} \tag{5.24}$$

These common characteristics are particularly useful when calculating state probabilities and state transition frequencies. The next example shows how this is used to calculate the expected time between load curtailment in a power system.

Example 5.5: State Enumeration and Markov Models

Question Consider the state enumeration of Example 5.2. If the unavailabilities and failure frequencies of the overhead lines are as shown in Table 5.3, how often does load curtailment occur?

Answer From the state enumeration of Example 5.2 we know that load curtailment only occurs when OHL1 & OHL3 fail, or when OHL2 & OHL4 fail. This is the case for $1000\,h/y$, while the rest of the year there are no first- or second-order contingencies that lead to load curtailment. In the first case, a failure of OHL1 & OHL3 can happen in two ways: first OHL1 then OHL3, or first OHL3 then OHL1. The probabilities of having either a failure of OHL1 or OHL3 is the system are:

$$P[\text{OHL1 failed}] = P_{S_{\text{OHL1}}} = U_1 A_2 A_3 A_4 = 2.0 \cdot 10^{-5} \qquad (5.25)$$

$$P[\text{OHL3 failed}] = P_{S_{\text{OHL3}}} = A_1 A_2 U_3 A_4 = 6.0 \cdot 10^{-5} \qquad (5.26)$$

Then, the state transition frequency from OHL1 failed (S_{OHL1}) to both OHL1 & OHL3 failed ($S_{\text{OHL1,OHL3}}$) is:

$$f_{S_{\text{OHL1}} \to S_{\text{OHL1,OHL3}}} = P_{S_{\text{OHL1}}} f_3 = 2.0 \cdot 10^{-5} \cdot 0.066 = 1.3 \cdot 10^{-6}\,/y \qquad (5.27)$$

And the state transition frequency from only OHL3 failed to both OHL1 & OHL3 failed is:

$$f_{S_{\text{OHL3}} \to S_{\text{OHL1,OHL3}}} = P_{S_{\text{OHL3}}} f_1 = 6.0 \cdot 10^{-5} \cdot 0.022 = 1.3 \cdot 10^{-6}\,/y \qquad (5.28)$$

Then, the total frequency of both OHL1 & OHL3 failed is:

$$\begin{aligned} f_{S_{\text{OHL1,OHL3}}} &= f_{S_{\text{OHL1}} \to S_{\text{OHL1,OHL3}}} + f_{S_{\text{OHL3}} \to S_{\text{OHL1,OHL3}}} \\ &= 1.3 \cdot 10^{-6} + 1.3 \cdot 10^{-6} = 2.6 \cdot 10^{-6}\,/y \end{aligned} \qquad (5.29)$$

Table 5.3 Unavailabilities and failure frequencies of the four overhead lines (Example 5.5)

Overhead line no.	OHL1	OHL2	OHL3	OHL4
Unavailability (U_n) [–]	$2 \cdot 10^{-5}$	$4 \cdot 10^{-5}$	$6 \cdot 10^{-5}$	$8 \cdot 10^{-5}$
Failure frequency (f_n) [–]	0.022	0.044	0.066	0.088

Strictly spoken this is an approximation, albeit an accurate one, as we do not include the state transitions from higher-order failure states back to state $S_{OHL1,OHL3}$. Similarly, for the failure of OHL2 & OHL4:

$$P[\text{OHL2 failed}] = P_{S_{OHL2}} = A_1 U_2 A_3 A_4 = 4.0 \cdot 10^{-5} \qquad (5.30)$$

$$P[\text{OHL4 failed}] = P_{S_{OHL4}} = A_1 A_2 A_3 U_4 = 8.0 \cdot 10^{-5} \qquad (5.31)$$

$$f_{S_{OHL2} \to S_{OHL2,OHL4}} = P_{S_{OHL2}} f_4 = 4.0 \cdot 10^{-5} \cdot 0.088 = 3.5 \cdot 10^{-6} \,/\text{y} \qquad (5.32)$$

$$f_{S_{OHL4} \to S_{OHL2,OHL4}} = P_{S_{OHL4}} f_2 = 8.0 \cdot 10^{-5} \cdot 0.044 = 3.5 \cdot 10^{-6} \,/\text{y} \qquad (5.33)$$

$$f_{S_{OHL2,OHL4}} = f_{S_{OHL2} \to S_{OHL2,OHL4}} + f_{S_{OHL4} \to S_{OHL2,OHL4}}$$

$$= 3.5 \cdot 10^{-6} + 3.5 \cdot 10^{-6} = 7.0 \cdot 10^{-6} \,/\text{y} \qquad (5.34)$$

Then, the frequency that load curtailment occurs becomes (when we take into account that this situation only applies for 1000 h/y):

$$f_{\text{ld curt}} = \frac{1000}{8760} \left(f_{S_{OHL1,OHL3}} + f_{S_{OHL2,OHL4}} \right)$$

$$= \frac{1000}{8760} \left(2.6 \cdot 10^{-6} + 7.0 \cdot 10^{-6} \right) = 1.1 \cdot 10^{-6} \,/\text{y} \qquad (5.35)$$

$$T_{\text{ld curt}} = \frac{1}{f_{\text{ld curt}}} = \frac{1}{1.1 \cdot 10^{-6}} = 9.1 \cdot 10^5 \,\text{y} \qquad (5.36)$$

This means that it is expected that load curtailment occurs once every 900,000 years, which indeed is very rare.

5.2 Generation Adequacy Analysis

In Sect. 3.4, it was explained how the stress-strength model can be used to model the reliability behavior of individual components. The stress-strength model can also be used effectively to model the reliability of complete generation systems consisting of many generators. In this case, the total load in the power system is seen as the system stress, while the strength of the generation system (i.e. all generators together) is seen as the system strength.

Fig. 5.6 Generation adequacy

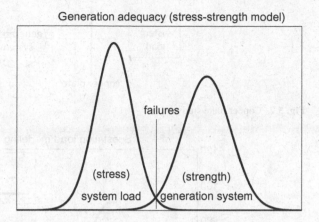

If the load level is higher than the total capacity of the available generators, the generation system fails to serve the load. This is illustrated in Fig. 5.6 (compare with Fig. 3.17). The area where the capacity of the generation system is smaller than the total system load is the area where the generation system fails.

In this model, normally a *copper plate approach* is used to model the power system, as illustrated in Fig. 5.7. This means that the network topology is completely neglected and modeled as one bus (the copper plate). To this bus, all loads and generators of the power system are connected. Because everything is connected to just one bus, the load can be aggregated to one system load and the generators can be aggregated to the generation system.

To apply the stress-strength model to generation adequacy, both the system load and the generation capacity must be described by probability distributions. The probability distribution of the load is mostly based on historical load levels, which can be combined with expected future load developments. In its simplest form, it can be a *normal distribution* (with mean and variance), but most domestic load follows a *combination of two normal distributions* (day versus night load levels). In Chap. 6, it is discussed how load can be modeled by so-called Gaussian Mixture Models. Also, *a histogram based on historical load values* or a *time series of the load* (e.g. hourly load values for one year) can be used. Figure 5.8 illustrates the modeling of a typical substation load.

5.2.1 Capacity Outage Probability Tables

The generation system must be represented by a probability distribution as well. This can be done by constructing a *Capacity Outage Probability Table (COPT)* of the generation system [2]. A COPT is a table that gives an overview of the possible generation capacity outages and the corresponding probabilities. The basic input information are the capacities of the generators and their (forced) unavailabilities.

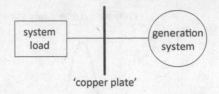

Fig. 5.7 Copper plate study

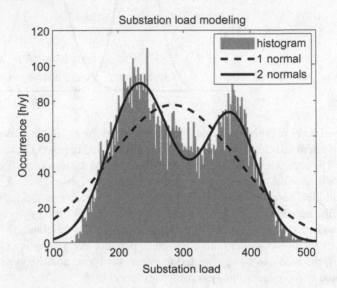

Fig. 5.8 Modeling of a substation load

In the past, these forced generator unavailabilities were often called *Forced Outage Rates (FOR)*. This is however confusing, as a rate is (like a frequency) measured in the unit [/y]. (Forced) unavailability therefore is a more suitable concept. It must be realized that there is a variety of possible generator failures, with effects ranging from immediate shutdown to the need to perform maintenance within several weeks. In this section, only generator failures that lead to immediate shutdown are considered, which is meant by (forced) unavailability.

A COPT can be created by state enumeration of the generator outages. Suppose we have two generators: one with capacity C_{g1} and (forced) unavailability U_{g1}, the other with capacity C_{g2} and (forced) unavailability U_{g2}. The capacity of the second generator is smaller than that of the first ($C_{g2} < C_{g1}$). If there are no generator failures, there is no capacity outage, and the probability of this situation is:

$$P[C_{\text{out}} = 0] = A_{g1}A_{g2} \tag{5.37}$$

If generator 1 has failed, the capacity outage is C_{g1}, with probability:

$$P[C_{\text{out}} = C_{g1}] = U_{g1}A_{g2} \tag{5.38}$$

If generator 2 has failed, the capacity outage is C_{g2}, with probability:

$$P[C_{\text{out}} = C_{g2}] = A_{g1}U_{g2} \tag{5.39}$$

And if both generators fail, the capacity outage is $C_{g1} + C_{g2}$, with probability:

$$P[C_{\text{out}} = C_{g1} + C_{g2}] = U_{g1}U_{g2} \tag{5.40}$$

The COPT then becomes like shown in Table 5.4.

A third column is now added to the COPT which shows 1 minus the cumulative probability of the capacity outages. It starts with 1, and for each following cell within this column, the probability in the cell on the left is subtracted. The result is shown in Table 5.5. It can be verified that the second column adds up to 1, while the values of the lowest cells of the second and third column are the same.

This process will become more clear in the following example. For larger generation systems, it is helpful to create the COPT using computer software like MATLAB or Microsoft Excel.

Example 5.6: COPT of a 4-Generator System

Question Assume we have a generation system consisting of four generators, with capacities and (forced) unavailabilities as specified in Table 5.6. Create the COPT of this generation system.

Table 5.4 COPT for the two generators

C_{out} [MW]	P [-]
0	$A_{g1}A_{g2}$
C_{g2}	$A_{g1}U_{g2}$
C_{g1}	$U_{g1}A_{g2}$
$C_{g1} + C_{g2}$	$U_{g1}U_{g2}$

Table 5.5 COPT for the two generators (with third column)

C_{out} [MW]	P [-]	$1 - \sum_{1}^{n-1} P$ [-]
0	$A_{g1}A_{g2}$	1
C_{g2}	$A_{g1}U_{g2}$	$1 - A_{g1}A_{g2}$
C_{g1}	$U_{g1}A_{g2}$	$1 - A_{g1}A_{g2} - A_{g1}U_{g2}$
$C_{g1} + C_{g2}$	$U_{g1}U_{g2}$	$1 - A_{g1}A_{g2} - A_{g1}U_{g2} - U_{g1}A_{g2}$

Table 5.6 Unavailabilities of the four generators (Example 5.6)

Capacity [MW]	Unavailability [–]
$C_{g1} = 200$	$U_{g1} = 0.05$
$C_{g2} = 100$	$U_{g2} = 0.03$
$C_{g3} = 200$	$U_{g3} = 0.04$
$C_{g4} = 500$	$U_{g4} = 0.06$

Table 5.7 COPT of the four-generator system (Example 5.6)

C_{out} [MW]	P [–]	$1 - \sum_1^{n-1} P$ [–]
0	0.8316	1
100	$2.6 \cdot 10^{-2}$	0.17
200	$7.8 \cdot 10^{-2}$	0.14
300	$2.4 \cdot 10^{-3}$	$6.4 \cdot 10^{-2}$
400	$1.8 \cdot 10^{-3}$	$6.2 \cdot 10^{-2}$
500	$5.3 \cdot 10^{-2}$	$6.0 \cdot 10^{-2}$
600	$1.6 \cdot 10^{-3}$	$6.9 \cdot 10^{-3}$
700	$5.0 \cdot 10^{-3}$	$5.3 \cdot 10^{-3}$
800	$1.5 \cdot 10^{-4}$	$2.7 \cdot 10^{-4}$
900	$1.2 \cdot 10^{-4}$	$1.2 \cdot 10^{-4}$
1000	$3.6 \cdot 10^{-6}$	$3.6 \cdot 10^{-6}$

Answer The probabilities of the generation capacity outages can be calculated by state enumeration of the possible generator outages. Please note that although the values in Table 5.7 are rounded to two digits, the calculation was made in a spreadsheet to avoid rounding errors. The probability of having 0 capacity outage can be calculated by:

$$P[C_{out} = 0] = \prod_{n=1}^{4} A_{gn} = A_{g1}A_{g2}A_{g3}A_{g4} = 0.8316 \qquad (5.41)$$

The probabilities of capacity outages caused by single generator failures are:

$$P[g1\ out, C_{out} = 200] = U_{g1}A_{g2}A_{g3}A_{g4} = 4.4 \cdot 10^{-2} \qquad (5.42)$$

$$P[g2\ out, C_{out} = 100] = A_{g1}U_{g2}A_{g3}A_{g4} = 2.6 \cdot 10^{-2} \qquad (5.43)$$

$$P[g3\ out, C_{out} = 200] = A_{g1}A_{g2}U_{g3}A_{g4} = 3.5 \cdot 10^{-2} \qquad (5.44)$$

$$P[g4\ out, C_{out} = 500] = A_{g1}A_{g2}A_{g3}U_{g4} = 5.3 \cdot 10^{-2} \qquad (5.45)$$

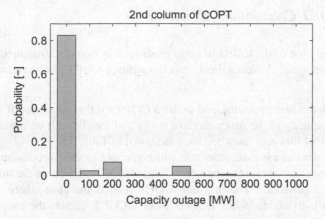

Fig. 5.9 Second column of the COPT of the four-generator system (Example 5.6)

The probabilities of capacity outages caused by two generator failures are:

$$P[\text{g1+g2 out}, C_{\text{out}} = 300] = U_{g1}U_{g2}A_{g3}A_{g4} = 1.4 \cdot 10^{-3} \qquad (5.46)$$

$$P[\text{g1+g3 out}, C_{\text{out}} = 400] = U_{g1}A_{g2}U_{g3}A_{g4} = 1.8 \cdot 10^{-3} \qquad (5.47)$$

etc.

After calculating the probability of all generator failure combinations, the probabilities of the same capacity outages can be added, like for example:

$$P[C_{\text{out}} = 200] = P[\text{g1 out}] + P[\text{g3 out}]$$
$$= 4.38 \cdot 10^{-2} + 3.46 \cdot 10^{-2} = 7.84 \cdot 10^{-2} \qquad (5.48)$$

The third column starts at 1 and for each next value, the probability in the cell to the left is subtracted. The complete COPT then becomes as shown in Table 5.7.

The second column of the COPT can also be plotted in a graph, which is shown Fig. 5.9. It can be seen that for these four generators, it is very likely to have no capacity outage. It is possible to have generation capacity outages equal to the capacities of the generators, while it is very unlikely to have capacity outages caused by multiple generator failures.

5.2.2 COPT Calculation Algorithm

For the calculation of the COPT of large generation systems, a structured algorithm has been developed [2]. According to this algorithm, a COPT can be created in a few steps:

1. Start with the first generator, and create a COPT for this generator. If we assume that the capacity of the first generator is C_{g1} and the (forced) unavailability U_{g1}, the COPT of this generator becomes as shown in Table 5.8.
2. Now we consider a second generator, with capacity C_{g2} and (forced) unavailability U_{g2}. We can now create two new (temporary) COPTs, based on the initial COPT of the first generator. The first new COPT reflects the case where the second generator is available, while the second new COPT reflects the case where the second generator is unavailable.

   The first new COPT can be created by multiplying the probabilities in the second column of the initial COPT by A_{g2}, because it reflects the case that the second generator is available. The capacity outages remain the same, because the second generator is available. The third column can be left empty.

   The second new COPT can be created by multiplying the probabilities in the second column of the initial COPT by U_{g2}, because it reflects the case that the second generator is unavailable. C_{g2} is added to the capacity outages in the first column, because the second generator is unavailable in this case. The third column can be left empty.

   The two new COPTs now become as shown in Table 5.9.
3. The two new COPTs are now merged into one COPT for the two generators. The two COPTs are merged in such a way that the capacity outages are increasing. If capacity outages happen to be the same, the probabilities of these capacity outage can be added. The third column starts with 1 and for each following cell within this column, the probability in the cell on the left is subtracted.

Table 5.8 COPT of a single generator

C_{out} [MW]	P [–]	$1 - \sum_1^{n-1} P$ [–]
0	A_{g1}	1
C_{g1}	U_{g1}	U_{g1}

Table 5.9 Two new COPTs for the second generator

C_{out} [MW]	P [–]	$1 - \sum_1^{n-1} P$ [–]
0	$A_{g1}A_{g2}$	
C_{g1}	$U_{g1}A_{g2}$	
C_{out} [MW]	P [–]	$1 - \sum_1^{n-1} P$ [–]
C_{g2}	$A_{g1}U_{g2}$	
$C_{g1} + C_{g2}$	$U_{g1}U_{g2}$	

Table 5.10 Combined COPT for the two generators

C_{out} [MW]	P [–]	$1 - \sum_1^{n-1} P$ [–]
0	$A_{g1}A_{g2}$	1
C_{g2}	$A_{g1}U_{g2}$	$1 - A_{g1}A_{g2}$
C_{g1}	$U_{g1}A_{g2}$	$1 - A_{g1}A_{g2} - A_{g1}U_{g2}$
$C_{g1} + C_{g2}$	$U_{g1}U_{g2}$	$1 - A_{g1}A_{g2} - A_{g1}U_{g2} - U_{g1}A_{g2}$

Assuming that $C_{g2} < C_{g1}$, the combined COPT becomes as shown in Table 5.10. It can be verified that the second column adds up to 1, while the values of the lowest cells of the second and third column are the same.

4. Steps 2 and 3 are now repeated for each next generator, until all generators of the generation system are included in the COPT. Care must be taken when combining the two temporary COPTs. The third column can be left empty until the final COPT is created. This whole process will become more clear in the following example.

Example 5.7: COPT Calculation Algorithm

Question Assume we have the generation system as described in the previous example (Example 5.6). Create the COPT using the described algorithm.

Answer The COPT of this generation system can be created in several steps:

1. The initial COPT, shown in Table 5.11, is based on the first generator. Please note that although the values in the tables are rounded to two digits, the calculation was made in a Microsoft Excel sheet to avoid rounding errors.
2. The second generator is now considered, leading to the temporary COPTs as shown in Table 5.12 (where the third columns are omitted).
3. Both temporary COPTs are now combined into one and the third column can be calculated, as shown in Table 5.13.
4. Now, the third generator is considered and the two new temporary COPTs as shown in Table 5.14 are created. It is decided to omit the third column until the final COPT is created.
5. These temporary COPTs are now combined into the COPT of Table 5.15.
6. Now, the fourth generator is considered and again two temporary COPTs are created, as shown in Table 5.16.
7. Again, the temporary COPTs are combined, where the probabilities of having 500 MW capacity outage are added. As this is the final COPT, the third column is added as well. Table 5.17 shows the final COPT.

Table 5.11 Initial COPT of the generation system (Example 5.7)

C_{out} [MW]	P [–]	$1 - \sum_1^{n-1} P$ [–]
0	0.95	1
200	0.05	0.05

Table 5.12 Two new (temporary) COPTs (Example 5.7)

C_{out} [MW]	P [–]
0	$0.95 \cdot 0.97 = 0.92$
200	$0.05 \cdot 0.97 = 4.9 \cdot 10^{-2}$
C_{out} [MW]	P [–]
100	$0.95 \cdot 0.03 = 2.9 \cdot 10^{-2}$
300	$0.05 \cdot 0.03 = 1.5 \cdot 10^{-3}$

Table 5.13 Combined COPT of the generation system (Example 5.7)

C_{out} [MW]	P [–]	$1 - \sum_1^{n-1} P$ [–]
0	0.92	1
100	$2.9 \cdot 10^{-2}$	$7.9 \cdot 10^{-2}$
200	$4.9 \cdot 10^{-2}$	$5.0 \cdot 10^{-2}$
300	$1.5 \cdot 10^{-3}$	$1.5 \cdot 10^{-3}$

Table 5.14 Two new (temporary) COPTs (Example 5.7)

C_{out} [MW]	P [–]
0	$0.92 \cdot 0.96 = 0.88$
100	$2.9 \cdot 10^{-2} \cdot 0.96 = 2.7 \cdot 10^{-2}$
200	$4.9 \cdot 10^{-2} \cdot 0.96 = 4.7 \cdot 10^{-2}$
300	$1.5 \cdot 10^{-3} \cdot 0.96 = 1.4 \cdot 10^{-3}$
C_{out} [MW]	P [–]
200	$0.92 \cdot 0.04 = 3.7 \cdot 10^{-2}$
300	$2.9 \cdot 10^{-2} \cdot 0.04 = 1.1 \cdot 10^{-3}$
400	$4.9 \cdot 10^{-2} \cdot 0.04 = 1.9 \cdot 10^{-3}$
500	$1.5 \cdot 10^{-3} \cdot 0.04 = 6.0 \cdot 10^{-5}$

Table 5.15 Combined COPT of the generation system (Example 5.7)

C_{out} [MW]	P [−]
0	0.8846
100	$2.7 \cdot 10^{-2}$
200	$4.7 \cdot 10^{-2} + 3.7 \cdot 10^{-2} = 8.3 \cdot 10^{-2}$
300	$1.4 \cdot 10^{-3} + 1.1 \cdot 10^{-3} = 2.6 \cdot 10^{-3}$
400	$1.9 \cdot 10^{-3}$
500	$6.0 \cdot 10^{-5}$

Table 5.16 Two new (temporary) COPTs (Example 5.7)

C_{out} [MW]	P [−]
0	$0.8846 \cdot 0.94 = 0.8316$
100	$2.7 \cdot 10^{-2} \cdot 0.94 = 2.6 \cdot 10^{-2}$
200	$8.3 \cdot 10^{-2} \cdot 0.94 = 7.8 \cdot 10^{-2}$
300	$2.6 \cdot 10^{-3} \cdot 0.94 = 2.4 \cdot 10^{-3}$
400	$1.9 \cdot 10^{-3} \cdot 0.94 = 1.8 \cdot 10^{-3}$
500	$6.0 \cdot 10^{-5} \cdot 0.94 = 5.6 \cdot 10^{-5}$
C_{out} [MW]	P [−]
500	$0.8846 \cdot 0.06 = 5.3 \cdot 10^{-2}$
600	$2.7 \cdot 10^{-2} \cdot 0.06 = 1.6 \cdot 10^{-3}$
700	$8.3 \cdot 10^{-2} \cdot 0.06 = 5.0 \cdot 10^{-3}$
800	$2.6 \cdot 10^{-3} \cdot 0.06 = 1.5 \cdot 10^{-4}$
900	$1.9 \cdot 10^{-3} \cdot 0.06 = 1.2 \cdot 10^{-4}$
1000	$6.0 \cdot 10^{-5} \cdot 0.06 = 3.6 \cdot 10^{-6}$

Table 5.17 Combined COPT of the generation system (Example 5.7)

C_{out} [MW]	P [−]	$1 - \sum_{1}^{n-1} P$ [−]
0	0.8316	1
100	$2.6 \cdot 10^{-2}$	0.17
200	$7.8 \cdot 10^{-2}$	0.14
300	$2.4 \cdot 10^{-3}$	$6.4 \cdot 10^{-2}$
400	$1.8 \cdot 10^{-3}$	$6.2 \cdot 10^{-2}$
500	$5.3 \cdot 10^{-2}$	$6.0 \cdot 10^{-2}$
600	$1.6 \cdot 10^{-3}$	$6.9 \cdot 10^{-3}$
700	$5.0 \cdot 10^{-3}$	$5.3 \cdot 10^{-3}$
800	$1.5 \cdot 10^{-4}$	$2.7 \cdot 10^{-4}$
900	$1.2 \cdot 10^{-4}$	$1.2 \cdot 10^{-4}$
1000	$3.6 \cdot 10^{-6}$	$3.6 \cdot 10^{-6}$

5.2.3   Loss of Load and Loss of Energy Indices

With the COPT, probabilistic indicators known as the *loss of load and loss of energy indices* can be calculated to indicate the reliability of the generation system [2]. Some of these indicators are:

- *LOLP*: *Loss Of Load Probability* [–]
  The probability that the demanded power cannot be supplied (partially or completely) by the generation system. The *LOLP* is often determined based on a per-hour study for a studied time period (usually a year).

$$LOLP = P[C < L] = \frac{\sum_{i=1}^{n} P_i[C_i < L_i]}{n} \tag{5.49}$$

where
$P[C < L]$	= probability that the generation capacity is smaller than the load [–]
$P_i[C_i < L_i]$	= probability that the gen. capacity is smaller than the load, hour i [–]
C	= total available generation capacity [MW]
C_i	= total available generation capacity for hour i [MW]
L	= total load level [MW]
L_i	= total load level for hour i [MW]
n	= total time of the studied period (8760 h for a whole year) [h]

- *LOLE*: *Loss Of Load Expectation* [h/y]
  Expected amount of time per period that the demanded power cannot be supplied (partially or completely) by the generation system.

$$LOLE = \sum_{i=1}^{n} P_i[C_i < L_i] = n \cdot LOLP \tag{5.50}$$

where
$P_i[C_i < L_i]$	= probability that the gen. capacity is smaller than the load, hour i [–]
C_i	= total available generation capacity for hour i [MW]
L_i	= total load level for hour i [MW]
n	= total time of the studied period (8760 h for a whole year) [h]
$LOLP$	= loss of load probability [–]

- *LOEE*: *Loss Of Energy Expectation* [MWh/y]
  Total amount of energy that is expected not to be supplied during a given time period (usually on a yearly basis) because of failures of the generation system.

$$LOEE = \sum_{i=1}^{n} \sum_{j \in S_g} P_j L_j \tag{5.51}$$

where
P_j = probability of generation capacity outage j [–]
L_j = not delivered load because of capacity outage j [MW]
n = total time of the studied period (8760 h for a whole year) [h]
S_g = set of possible generation capacity outages

Example 5.8: Loss-of-Load and Loss-of-Energy Indices

Question Consider the generation system as described in Example 5.6 (and 5.7). Calculate the *LOLP*, *LOLE* and *LOEE* if the total system load is 550 MW for 5000 h/y, and 350 MW 3760 h/y.

Answer We start with calculating the *LOLP*. To avoid confusion, we will make a distinction in notation between available generation capacity C_{av} and generation capacity outage C_{out}. We first calculate the probability that the available generation capacity is smaller than the load ($P[C_{\text{av}} < L]$). The total installed capacity of the generation system is 1000 MW. If the load is 550 MW, a generation capacity outage larger than 450 MW (1000–550) will result in loss of load. The third column of the final COPT in Example 5.6 (Table 5.7) shows that the probability of having a generation capacity outage $\geq$500 MW is $6.0 \cdot 10^{-2}$ (a capacity outage of 450 MW is not possible because of the generator sizes):

$$P[C_{\text{av}} < 550] = P[C_{\text{out}} > 450] = P[C_{\text{out}} \geq 500] = 6.0 \cdot 10^{-2} \quad (5.52)$$

If the load is 350 MW, a generation capacity outage larger than 650 MW (1000–350) will result in loss of load. The third column of the COPT of Table 5.7 shows that the probability of a generation capacity outage $\geq$ 700 MW is $5.3 \cdot 10^{-3}$ (a capacity outage of 650 MW is not possible because of the generator sizes):

$$P[C_{\text{av}} < 350] = P[C_{\text{out}} > 650] = P[C_{\text{out}} \geq 700] = 5.3 \cdot 10^{-3} \quad (5.53)$$

Now, we can combine this to calculate the *LOLP*:

$$LOLP = \frac{\sum_{i=1}^{n} P_i[C_i < L_i]}{n} = \frac{5000 \cdot P[C_{\text{av}} < 550] + 3760 \cdot P[C_{\text{av}} < 350]}{8760}$$
$$= \frac{5000 \cdot 6.0 \cdot 10^{-2} + 3760 \cdot 5.3 \cdot 10^{-3}}{8760} = 3.65 \cdot 10^{-2} \quad (5.54)$$

The *LOLE* can be calculated directly from the *LOLP*:

$$LOLE = n \cdot LOLP = 8760 \cdot 3.65 \cdot 10^{-2} = 320 \, \text{h/y} \quad (5.55)$$

Table 5.18 COPT of the generation system with the capacity of not-delivered load (Example 5.8)

C_{out} [MW]	P_j [–]	L_j [MW] (550 MW load)	L_j [MW] (350 MW load)
0	0.8316	0	0
100	$2.6 \cdot 10^{-2}$	0	0
200	$7.8 \cdot 10^{-2}$	0	0
300	$2.4 \cdot 10^{-3}$	0	0
400	$1.8 \cdot 10^{-3}$	0	0
500	$5.3 \cdot 10^{-2}$	50	0
600	$1.6 \cdot 10^{-3}$	150	0
700	$5.0 \cdot 10^{-3}$	250	50
800	$1.5 \cdot 10^{-4}$	350	150
900	$1.2 \cdot 10^{-4}$	450	250
1000	$3.6 \cdot 10^{-6}$	550	350

To calculate the *LOEE* we will use the second column of the COPT of Table 5.7, as we need to know the amount of energy that cannot be supplied for every possible capacity outage. It may be helpful to make a table with capacity outages, probabilities and not-delivered load, such as shown in Table 5.18. For example, if the total system load is 550 MW, a generation capacity outage of 600 MW (i.e. 400 MW generation capacity still available) leads to 150 MW lost load. And if the total system load is 350 MW, a generation capacity outage of 800 MW (i.e. 200 MW generation capacity still available) leads to 150 MW lost load as well.

Two columns can be added to this table. One in which the values of the second and third column are multiplied, and one in which the values of the second and fourth column are multiplied. The sum of these columns is calculated as well. Table 5.19 shows the result of this.

The *LOEE* can now be calculated as:

$$LOEE = \sum_{i=1}^{n} \sum_{j \in S_g} P_j L_j = 5000 \sum_{j \in S_g} P_j L_{j,(550Ld)} + 3760 \sum_{j \in S_g} P_j L_{j,(350Ld)}$$

$$= 5000 \cdot 4.25 + 3760 \cdot 0.30 = 22 \cdot 10^3 \text{ MWh/y} = 22 \text{ GWh/y} \quad (5.56)$$

After this example, it may be clear that the calculation of the COPT and related reliability indicators can best be done using a computer program, especially if the total system load consists of 8760 different values for a year.

Table 5.19 COPT of the generation system with the amount of lost load (Example 5.8)

C_{out}	P_j	$L_{j,(550Ld)}$	$P_j L_{j,(550Ld)}$	$L_{j,(350Ld)}$	$P_j L_{j,(350Ld)}$
0	0.8316	0	0	0	0
100	$2.6 \cdot 10^{-2}$	0	0	0	0
200	$7.8 \cdot 10^{-2}$	0	0	0	0
300	$2.4 \cdot 10^{-3}$	0	0	0	0
400	$1.8 \cdot 10^{-3}$	0	0	0	0
500	$5.3 \cdot 10^{-2}$	50	2.7	0	0
600	$1.6 \cdot 10^{-3}$	150	0.24	0	0
700	$5.0 \cdot 10^{-3}$	250	1.3	50	0.25
800	$1.5 \cdot 10^{-4}$	350	0.053	150	0.023
900	$1.2 \cdot 10^{-4}$	450	0.054	250	0.030
1000	$3.6 \cdot 10^{-6}$	550	0.0020	350	0.0013
Sum			4.25		0.30

Table 5.20 Overview of the generation system of the Netherlands (Example 5.9)

Capacity [MW]	Units [–]	$\sum C_g$ [MW]	C_{mean} [MW]
$C_g \leq 100$	24	1117	46.5
$100 < C_g \leq 250$	12	2209	184
$250 < C_g \leq 500$	24	8986	374
$500 < C_g$	25	17919	717
Total	85	30231	356

Example 5.9: Generation System Adequacy Netherlands

In [6], the reliability of the Dutch generation system and the effects of increasing installed offshore wind capacity were investigated. The objective was to study the impact of increasing offshore wind capacity and different offshore grid topologies on the reliability of the power system. As part of this research, a generation adequacy study was performed which included all generators in the Dutch power system. An overview of the generators is shown in Table 5.20.

In reality, the number of small (<100 MW) units is much larger than mentioned in the table, but these small units were considered as distributed generation and excluded from the generation adequacy study. The (forced) unavailabilities of the generators were estimated and ranged from 0.04–0.09, with an average of 0.077. From this data, a COPT of the Dutch generation system was created. The second column of this COPT is illustrated in Fig. 5.10.

The graph shows that the COPT follows a somewhat skewed normal distribution. The distribution is skewed because capacity outages smaller than 0 are not possible. Furthermore, some peaks can be seen on the left, which are

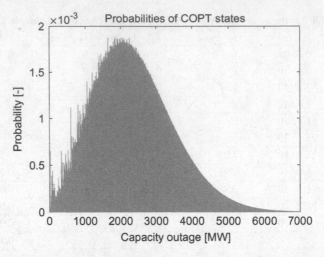

Fig. 5.10   Probability of the COPT states of the Dutch generation system (Example 5.9)

the failure probabilities of single (and some combinations of two) generators. The smooth tail on the right is caused by combinations of multiple generator failures. It is also interesting that the peak of the probability distribution is around 2000 MW. In a generation system of this size, there are always some generators unavailable.

Compared to the system strength of the stress-strength model (Fig. 5.6), the graph of the capacity outage probabilities is mirrored. A COPT shows generation capacity outages, while the generation system strength is the available generation capacity. If the capacity outages are subtracted from the total installed generation capacity, the generation system strength is obtained. In this study, a year scenario was used for the system load, consisting of hourly values for the year 2020. A histogram of the load was made and plotted together with the available generation capacity in one graph, as illustrated in Fig. 5.11. The similarities with Fig. 5.6 are clear now. There is a small area in which the load is larger than the available generation capacity.

Using the load scenario and the COPT, the reliability indices *LOLE* and *LOEE* were calculated. In total, a *LOLE* of $10.67 \cdot 10^{-7}$ h/y and a *LOEE* of $3.29 \cdot 10^{-4}$ MWh/y were found. The generation system is thus very reliable. The graphs in Fig. 5.12 show how the *LOLE* and *LOEE* can be divided over the hours of the year. From these graphs, it can be concluded that the *LOLE* and *LOEE* are mainly determined by a few hours of the year with the highest total system load.

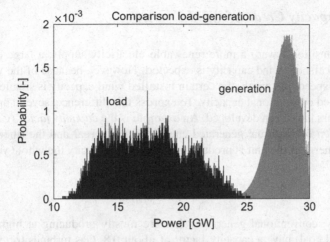

Fig. 5.11 Comparison of the COPT states and the load of the Dutch power system (Example 5.9)

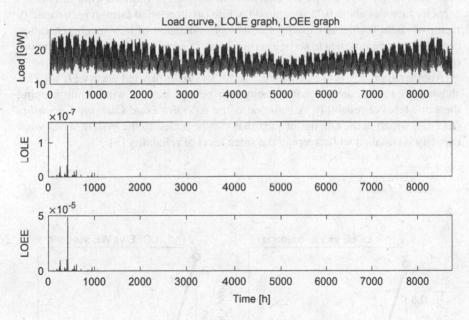

Fig. 5.12 *LOLE* and *LOEE* for the Dutch power system (Example 5.9)

5.2.4   Capacity Credit of Wind Energy

With the transition toward a more renewable electricity supply, a large expansion of installed offshore wind capacity is expected. However, because of the variability of the wind speed, the worth of a certain installed wind capacity is smaller than the same installed conventional capacity. To express this difference, several approaches and indicators have been developed. An example is the *capacity factor (CF)*, which is the fraction between the generated energy within a year and the (theoretically) generated energy if the unit is producing at installed capacity the whole year:

$$CF = \frac{E_{\text{generated}}}{8760 C_{\text{installed}}} \tag{5.57}$$

Normally, conventional generators that are mostly producing at high capacity (nuclear and coal) have a capacity factor of about 0.8. Gas turbines (often used as peak-load units) have a capacity factor of about 0.4. For onshore wind turbines, a capacity factor of about 0.35 is common, while offshore wind farms have a capacity factor of about 0.45. The capacity factor is an indication of how much generation capacity must be installed to serve a certain load. It, however, focuses on the amount of demanded energy and not on whether the energy is generated at the right time.

Another approach is to calculate the *capacity credit* of wind energy [7, 8]. This shows how much conventional capacity can be replaced by wind while keeping the same level of reliability. A variation is the *Effective Load Carrying Capability (ELCC)*, which is the amount of load that can be added to the system when wind capacity is installed while keeping the same level of reliability [8].

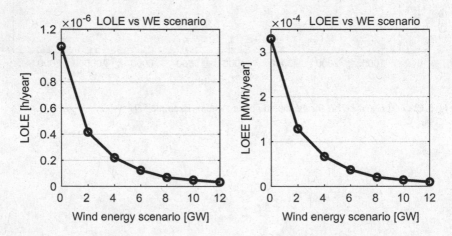

Fig. 5.13 *LOLE* and *LOEE* for increasing installed wind capacity (Example 5.10)

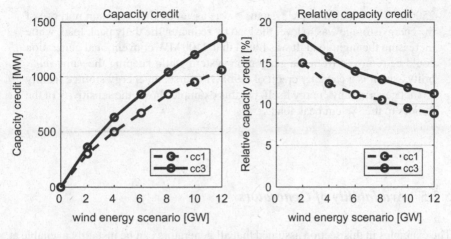

Fig. 5.14 (Relative) capacity credit for increasing installed wind capacity (Example 5.10)

Example 5.10: Capacity Credit of Wind in the Netherlands

The research as described in Example 5.9 was continued by increasing the installed offshore wind capacity [6]. For increasing offshore wind capacity, the reliability indices *LOLE* and *LOEE* were calculated. The result is shown in the graphs of Fig. 5.13. It can be seen that these indices reduce when more wind capacity is included. Logically, installing more generation capacity leads to a more reliable generation system. The increase in reliability is large for the first GWs of offshore wind, while the effect of more wind capacity becomes smaller.

It was now studied how much conventional generation can be replaced by offshore wind. The graphs in Fig. 5.14 show the Capacity Credit (cc3) and Effective Load Carrying Capability (cc1) for increasing offshore wind capacities. The graph on the left shows how much capacity can be replaced by offshore wind. The graph on the right is the most important and shows the capacity credit and *ELCC* as a percentage. It can be concluded that the capacity credit of offshore wind ranges from 0.11–0.18, which is actually a very small value.

The benefit of this approach is that is focuses on the reliability of the generation system, thereby including the variable nature of wind energy. A drawback is that it is directly related to reliability indices like *LOLE* and *LOEE*, which are mainly determined by the peak load of the system. In other words, it is strongly dependent on how much the wind capacity can reduce the peak load of the system.

In this respect, it is interesting to calculate the capacity credit of energy storage. In the research, a storage facility with capacity 30 GWh and power

2500 MW was added to the system. It was assumed that the main purpose of the energy storage was to level the load by reducing the daily peak load, while increasing the night load. It was found that 2400 MW conventional generation could be removed from the generation system while keeping the same reliability, which is a capacity credit of 0.96. Of course, an energy storage facility cannot generate any energy itself, but this example shows the sensitivity of the indices to the system peak load.

5.2.5 *Availability of Generators*

The examples in this section assumed that all generators can be instantly available at their installed capacities. In reality, this is not the case. The described studies therefore mainly answer the question whether there is theoretically enough installed generation capacity to supply the load. For more detailed studies, it might be necessary to include maintenance and start-up times of generators. Some generators will be unavailable because of planned maintenance, while other generators (like coal power plants) might not be available within a short time. A COPT can then be made of the generators that can be available within the desired time interval. This COPT will vary throughout the year. Using these COPTs, a more accurate calculation of the generation system reliability can be performed corresponding specific conditions.

5.3 Monte Carlo Simulation

In some cases, it can be too complicated to analyze the reliability of a power system analytically. For example, there might be too many possible failure states or it is expected that higher-order failure states contribute significantly to the system (un)reliability. In other cases, it may be desired to collect more information, like probability distributions of the amount of load curtailment or one might wish to include real system behavior like switching actions of protection systems. In these cases, it is possible to simulate the system behavior in a Monte Carlo simulation.

Monte Carlo simulation is a (computer) simulation in which the system behavior is simulated and from which results like probabilistic reliability indicators are extracted. Because it is a simulation of the system, it is possible to include real system behavior like load/generation time series, switching actions and protection systems in the study. Monte Carlo simulation is applied to a wide variety of studies, not only power system reliability analysis. The following example illustrates the principle of Monte Carlo simulation in a simple study.

Fig. 5.15 Example of
Monte Carlo simulation to
determine the value of π
(Example 5.11)

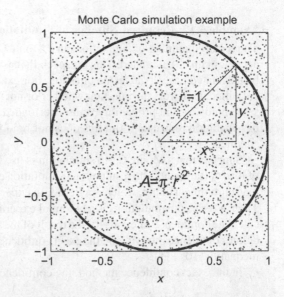

Example 5.11: Simple Monte Carlo Simulation Example

The principle of Monte Carlo simulation can be explained by a simple example
[9]. Suppose that we do not know the value of Pi (π). But we know that a circle
can be described by $x^2 + y^2 = r^2$ and that the area of the circle is $A = \pi r^2$.
Using Monte Carlo simulation we can estimate the value of π. First, a circle
with radius $r = 1$ is drawn, like in Fig. 5.15.

Then, two times 2500 random values are sampled from a uniform distribution between -1 and 1. These samples form x-values and y-values of points that
are drawn in the figure as well. It can now be determined how many of these
points are within the circle (i.e. $x^2 + y^2 < 1$). The number of points within the
circle (N_{circle}) divided by the total number of points (N_{square}) multiplied by 4,
i.e. the area of the square ($-1 \leq x \leq 1$; $-1 \leq y \leq 1$), is the estimate of π. In
this case:

$$\pi_{est} = 4 \frac{N_{circle}}{N_{square}} = 4 \cdot \frac{1939}{2500} = 3.10 \qquad (5.58)$$

One characteristic of Monte Carlo simulation is that it is less sensitive to the
complexity and dimension of the studied problem [9]. In fact, the accuracy of the
results is directly related to the number of simulations. The benefit is that studying
a more complex system does not require more simulations. But the disadvantage
is that the number of simulations mostly has to be very large in order to obtain a
reasonable accuracy of the results. This is the reason why Monte Carlo simulation is
often considered as a last resort.

Example 5.12: Required Number of Simulations

The required number of simulations in a Monte Carlo simulation for a desired accuracy can be calculated. This is especially easy if the reliability indicator to be calculated answers a yes/no question. For example, the probability of load curtailment answers the question whether or not there is load curtailment. The same holds for the probability of an overloaded network and the probability of generation redispatch. But this does not hold for the amount of energy not supplied.

Suppose that a Monte Carlo simulation is used to determine the probability of load curtailment. The result of one simulation can then either be that there is load curtailment or not, which is a Bernoulli experiment (see also Appendix A). Then, a series of simulations is a binomial experiment [10]. This enables us to calculate confidence intervals for the result of the simulation. Some methods to define confidence intervals are the exact confidence method and the Chebyshev inequality [10, 11].

In the exact confidence method, the confidence interval is defined as [11]:

$$p_{ub} = 1 - \text{BetaInv}\left(\frac{\alpha}{2}, n - k, k + 1\right) \tag{5.59}$$

$$p_{lb} = 1 - \text{BetaInv}\left(1 - \frac{\alpha}{2}, n - k + 1, k\right) \tag{5.60}$$

where

p_{ub}	= confidence interval upper bound [–]
p_{lb}	= confidence interval lower bound [–]
BetaInv()	= Beta Inverse function
n	= number of trials [–]
k	= number of successes in n trials [–]
α	= 1 - desired confidence level [–]

Suppose that the calculated *PLC* is 0.035 h/y ($p = 4 \cdot 10^{-6}$) and the number of simulations is $n = 4 \cdot 10^8$. Then:

$$k = pn = 4 \cdot 10^{-6} \cdot 4 \cdot 10^8 = 1.6 \cdot 10^3 \tag{5.61}$$

The confidence levels can be now calculated (e.g. by Microsoft Excel or Matlab) and become as shown in Table 5.21.

The required number of simulations can also be calculated by the Chebyshev inequality [10]. According to this Chebyshev inequality, the confidence interval is:

$$P\left[\left|\hat{P}_n(A) - P[A]\right| < c\right] \geq 1 - \frac{P[A](1 - P[A])}{nc^2} \tag{5.62}$$

Table 5.21 Confidence levels of various Monte Carlo simulations (Example 5.12)

Simulations (n)	$4 \cdot 10^8$	$4 \cdot 10^8$
Confidence level (α)	0.05 (95%)	0.01 (99%)
Upper confidence limit (p_{ub})	$4.2 \cdot 10^{-6}$ (0.0368 h/y)	$4.3 \cdot 10^{-6}$ (0.0373 h/y)
Probability event (p)	$4.0 \cdot 10^{-6}$ (0.0350 h/y)	$4.0 \cdot 10^{-6}$ (0.0350 h/y)
Lower confidence limit (p_{lb})	$3.8 \cdot 10^{-6}$ (0.0333 h/y)	$3.7 \cdot 10^{-6}$ (0.0328 h/y)

Table 5.22 Confidence levels of various Monte Carlo simulations (Example 5.12)

Confidence level (α)	95%	99%	95%	95%
Probability event (α)	$1 \cdot 10^{-4}$	$1 \cdot 10^{-4}$	0.50	0.50
Confidence interval (c)	$0.5 \cdot 10^{-4}$	$0.5 \cdot 10^{-4}$	$2.5 \cdot 10^{-3}$	$0.5 \cdot 10^{-4}$
Simulations (n)	$8.0 \cdot 10^5$	$4.0 \cdot 10^6$	$8.0 \cdot 10^5$	$2.0 \cdot 10^9$

where

$P\left[\left|\hat{P}_n(A) - P[A]\right| < c\right]$ = confidence level [–]

$P[A]$ = success probability of experiment [–]

$\hat{P}_n(A)$ = sample mean (estimation of $P[A]$) [–]

n = number of independent trials of an experiment [–]

c = half the confidence interval [–]

Now, the required number of simulations for a given confidence bound can be calculated as:

$$n = \frac{P[A](1 - P[A])}{\left(1 - P\left[\left|\hat{P}_n(A) - P[A]\right| < c\right]\right) c^2} \tag{5.63}$$

The required number of simulations for various confidence levels and event probabilities are shown in Table 5.22. It can be seen that the number of simulations for a confidence level of 95% and an event probability of 10^{-4} is almost one million. This value becomes even larger for a higher confidence level or a higher event probability.

5.3.1 Random Sampling of Component Failures

Monte Carlo simulation basically consists of three parts:

1. Sampling of component failures;
2. System simulation (e.g. load flow, remedial actions);
3. Processing of results.

In Monte Carlo simulation, a distinction can be made between *sequential* and *non-sequential* simulation. In a *non-sequential Monte Carlo simulation*, time is not considered. For each simulation instant, the load and generation are sampled from a load/generation probability distribution and the component failures are sampled using their unavailabilities. In a *sequential Monte Carlo simulation*, the time is considered. This means that the load and generation follow a time scenario, for example a year scenario of hourly values. For the component failures, the time to failure and time to repair are then sampled, mostly from the exponential distribution.

Sampling from the component (un)availabilities in a non-sequential Monte Carlo simulation is straightforward. If u is a sample from the unit uniform distribution, then a component n is assumed to be failed if $u < U_n$. In this way, the statuses of all components are determined. Sampling from the exponential distribution in a sequential Monte Carlo simulation is somewhat different. For the failure distribution:

$$F(t) = 1 - e^{-\lambda t} \tag{5.64}$$

And the inverse function:

$$F^{-1}(t) = \frac{-\ln(1 - F(t))}{\lambda} \tag{5.65}$$

If u is a sample from the unit uniform distribution, a sample from the exponential distribution is:

$$\frac{-\ln(1 - u)}{\lambda} \tag{5.66}$$

If λ is the failure rate in [/y], a sampled time to failure in [h] is (realizing that there is no significant difference between $(1 - u)$ and u):

$$TTF_{\text{sample}} = -8760 \frac{\ln(u)}{\lambda} \tag{5.67}$$

And if r is the average repair rate in [h] or μ is the repair rate in [/y], a sampled time to repair is:

$$TTR_{\text{sample}} = -8760 \frac{\ln(u)}{\mu} = -r \ln(u) \tag{5.68}$$

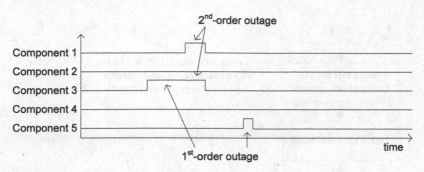

Fig. 5.16 Monte Carlo simulation

The sampling of times to failure and times to repair results in a time series that shows when components fail and are repaired, as illustrated in Fig. 5.16. This is used together with the load/generation scenario in the Monte Carlo simulation.

5.3.2 Monte Carlo Simulation Algorithm

A typical algorithm for (sequential) Monte Carlo simulation is shown in Fig. 5.17. As can be seen, the algorithm starts by sampling the first times to failure of the system components. It is assumed that all components are working when the simulation starts. The network model is then created, and the simulation starts. Because a long simulation time is required to obtain reasonable accuracy, it is common to simulate the same year multiple times. The load/generation scenario then follows a loop until the Monte Carlo simulation ends.

The core of the Monte Carlo simulation consists of the same parts as state enumeration. Load flow analysis is performed and if needed, remedial actions are applied. The simulation ends when the desired simulation time is reached. Every time a component fails or is repaired, the network model is created again and a new time to repair or time to failure is sampled for the specific component.

To speed up a Monte Carlo simulation, it is possible to not analyze the situations of which it is known that these do not cause any problems. For example, if it is known (e.g. by contingency analysis) that a power system is $N - 1$ redundant, analysis of the normal operation state and single contingencies can be skipped. This will speed up the simulation considerably as the simulation time must be very long to include rare higher-order contingencies as well.

During the Monte Carlo simulation, results from the remedial actions are collected. Reliability indicators are then calculated from these results. It is also possible to calculate the reliability indicators during the simulation to show the progress of the simulation.

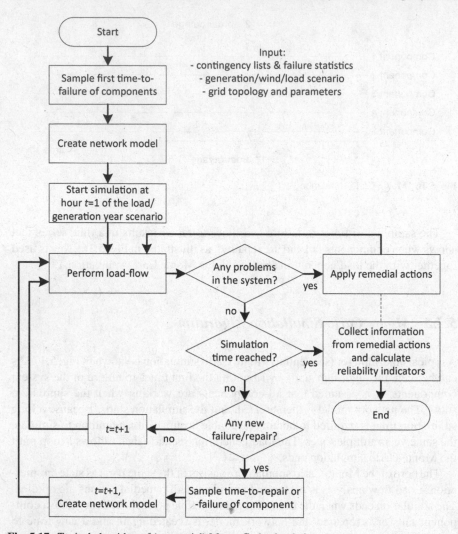

Fig. 5.17 Typical algorithm of (sequential) Monte Carlo simulation

Example 5.13: Monte Carlo Simulation Program

Question Design a software algorithm that performs Monte Carlo simulation. The requirements are:

- Perform a non-sequential Monte Carlo simulation of a transmission network.
- Simulate independent and dependent double contingencies, and the higher-order combinations.
- Sample load/generation from an hourly year scenario (8760 h).

- There are no system problems if there are no contingencies (normal operation)
- Perform a DC load flow and test whether the network is overloaded.
- Apply generation redispatch and load curtailment if needed.
- Calculate the probability of redispatch and the probability of load curtailment.
- Plot the intermediate results during the simulation.
- The simulation time is $t_s = 1$ million (simulated) hours.

Answer The precise implementation depends on the software that is used, but the general algorithm will be:

% initial values
$\text{cont}_{1\text{ind}}$ = list of single contingencies
$P_{1\text{ind}}$ = probabilities of single contingencies
$n_{1\text{ind}}$ = number of single contingencies
$\text{cont}_{2\text{dep}}$ = list of dependent double contingencies
$P_{2\text{dep}}$ = probabilities of double contingencies
$n_{2\text{dep}}$ = number of double contingencies
$t_{\text{sim}} = 0$ (this is the time of the simulation)
$P_{\text{redispatch}} = 0$ (this will be the probability of redispatch)
$P_{\text{ld.curt}} = 0$ (this will be the probability of load curtailment)
create an empty graph in which $P_{\text{redispatch}} = 0$ and $P_{\text{ld.curt}} = 0$ will be plotted

% Monte Carlo simulation
while($t_{\text{sim}} < 1 \cdot 10^6$)
    create a vector $u_{1\text{ind}}$ of $n_{1\text{ind}}$ unit uniform random samples
    $u_{1\text{ind}} < P_{1\text{ind}}$ gives a vector that indicates which single contingencies
    occur
    create a vector $u_{2\text{dep}}$ of $n_{2\text{dep}}$ unit uniform random samples
    $u_{2\text{dep}} < P_{2\text{dep}}$ gives a vector that indicates which double contingencies
    occur
    if(there are any contingencies)
        based on these contingencies, create the network model
        take a sample u from the unit uniform distribution
        the load/generation snapshot is hour ceil($8760 \cdot u$) from the scenario
        perform a dc load flow
        if(there is an overload in the network)
            $P_{\text{redispatch}} = P_{\text{redispatch}} + 1$
            perform generation redispatch
            if(there is still an overload in the network)
                perform load curtailment
                $P_{\text{ld.curt}} = P_{\text{ld.curt}} + 1$
            end if
        end if

```
        end if
        if(t_sim = 0, 1000, 2000, ..., 1 · 10^6)
            plot point (t_sim, P_redispatch/t_sim) in graph
            plot point (t_sim, P_ld.curt/t_sim) in graph
        end if
        t_sim = t_sim + 1
    end while

    % simulation results
    P_redispatch = P_redispatch/1 · 10^6
    P_ld.curt = P_ld.curt/1 · 10^6
```

Example 5.14: Monte Carlo Simulation Example

In a study, Monte Carlo simulation was used to verify the results of a state enumeration. The objective was to find the probability of an overloaded network and the probability of an islanded substation. The results of the Monte Carlo simulation plotted during the simulation are shown in Fig. 5.18. In the graphs, the probabilities have been scaled to [h/y]. It can be seen that the results slowly converge to the final value. To determine the accuracy of the results, confidence intervals according to Example 5.11 were calculated. In the graphs, it can be seen that the confidence intervals become smaller as the simulation continues. Still, the convergence is very weak. After a simulation time of $2 \cdot 10^7$ h, the real probability of an overloaded network still can be between 0.7–1.4 of the

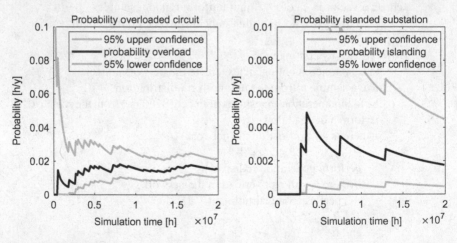

Fig. 5.18 Probability of an overloaded circuit and probability of an islanded substation in the example Monte Carlo simulation (Example 5.14)

estimated value (for a 95% confidence level). The probability of an islanded substation in this particular network is very small. The graph shows that the confidence interval decreases even more slowly than for the probability of an overloaded circuit. Generally, convergence of the result is slower for smaller probabilities.

5.4 Conclusion

In this chapter, methods to analyze the reliability of large systems were discussed. Generation adequacy analysis is used to study the reliability of generation systems. State enumeration and Monte Carlo simulation can be used to study the reliability of combined generation/transmission/distribution systems.

Both state enumeration and Monte Carlo simulation require a large amount of calculations. In state enumeration, the number of system states easily increases for larger systems and higher-order failures. Monte Carlo simulation is less sensitive to the dimensionality of the problem. However, Monte Carlo simulation requires a long simulation time to obtain a reasonable accuracy of the results. To reduce the number of calculations, various techniques can be used. For example, for Monte Carlo simulation, methods exist to increase the convergence speed [9]. In state enumeration, an intelligent choice of the states can reduce the amount of calculations needed.

The choice for a certain method always depends on the kind of study and the studied system. If only the generation system is considered, generation adequacy analysis can be performed. If higher-order failure states are expected to be of minor influence, state enumeration is a good approach. And if higher-order failure states are expected to be of influence or if the time behavior of the power system needs to be included, Monte Carlo simulation will probably be preferred.

Problems

5.1 Learning Objectives
Please check the learning objectives as given at the beginning of this chapter.

5.2 Main Concepts
Please check whether you understand the following concepts:

- State enumeration:
  - State enumeration;
  - Deterministic contingency analysis;

- $N - 1$ redundancy criterion;
- Probabilistic reliability indicators;
- Probability of Load Curtailment (*PLC*);
- Expected Energy Not Supplied (*EENS*);
- System Average Interruption Duration Index (*SAIDI*);
- System Average Interruption Frequency Index (*SAIFI*);
- Customer Average Interruption Duration Index (*CAIDI*);
- Probability of overload;
- Probability of generation redispatch;
- Expected redispatch costs;
- Probability of the alert state;
- Investment costs, operation costs, risk costs;
- (Net) Present Value, return on investment.

- Generation adequacy:

  - Generation adequacy analysis;
  - Copper plate approach;
  - Capacity Outage Probability Table (COPT);
  - Loss Of load and loss of energy indices;
  - Loss Of Load Probability (*LOLP*);
  - Loss Of Load Expectation (*LOLE*);
  - Loss Of Energy Expectation (*LOEE*);
  - Capacity credit (of wind energy);
  - Capacity factor (*CF*).

- Monte Carlo simulation:

  - Monte Carlo simulation;
  - Sequential and non-sequential simulation;
  - Random sampling of component failures.

5.3 State Enumeration
Explain how state enumeration works. How is it related to deterministic contingency analysis? Are there similarities with Markov models? In which cases is state enumeration preferred over Monte Carlo simulation?

5.4 State Enumeration Algorithm
Design a state enumeration algorithm (workflow and/or program code) that analyzes the reliability of a transmission network. The state enumeration must:

- Study independent single contingencies and combinations of independent single contingencies up to the third order.
- Perform an AC load flow for every hour of the year of a load/generation scenario and test whether there are any problems in the network (which could be overloading, over-/undervoltages).
- Apply generation redispatch and load curtailment if needed.
- Calculate the probability of generation redispatch and load curtailment.

• Calculate the error, in the sense of total probability of not-considered states.

5.5 Generation Adequacy Analysis

What is generation adequacy analysis? How is it related to the stress-strength model? How is it performed and what exactly is studied? Mention some probabilistic reliability indicators that can be calculated in generation adequacy analysis. How can the reliability of power systems with renewable sources like offshore wind be studied using generation adequacy?

5.6 Reliability Generation System

Assume we have a generation system consisting of only three generators. The capacities are $C_{g1} = 100\,\text{MW}$, $C_{g2} = 100\,\text{MW}$, and $C_{g3} = 200\,\text{MW}$. The generator unavailabilities are $U_{g1} = 0.03$, $U_{g2} = 0.04$, and $U_{g3} = 0.05$. Create the COPT of this generation system. Calculate the *LOLP* and *LOLE* if the load is always 190 MW. What are the *LOLP* and *LOLE* if the load is always 240 MW. Calculate the *LOLP* and *LOLE* if the load is 1/4 of the year 240 MW and 3/4 of the year 190 MW.

5.7 Monte Carlo Simulation

Describe what Monte Carlo simulation is and how it works. What do you know about simulation time, system complexity and the accuracy of the results? In which cases is Monte Carlo simulation preferred over state enumeration? What is the difference between sequential and non-sequential Monte Carlo simulation?

5.8 Monte Carlo Algorithm

Design a Monte Carlo simulation algorithm (workflow and/or program code) that analyzes the reliability of a transmission network. The requirements are:

• Perform a non-sequential Monte Carlo simulation.
• Simulate independent and dependent double contingencies, and the higher-order combinations.
• Sample load/generation from an hourly year scenario (8760 h).
• Perform an AC load flow and test whether there are any problems in the network (which could be overloading, over-/undervoltages).
• Apply generation redispatch and load curtailment if needed.
• Calculate the probability and amount of redispatch and load curtailment.
• The simulation time is $t_s = 1 \cdot 10^6$ (simulated) hours.

5.9 Probabilistic Reliability Indicators

Mention some probabilistic reliability indicators that are used to indicate the reliability of large systems. What exactly do they indicate? Which indicators are directly related to security of supply?

References

1. Kandalepa N, Tuinema BW, Rueda JL, Meijden MAMM van der (2015) Reliability modeling of transmission networks: an explanatory study on further EHV underground cabling in the Netherlands, MSc Thesis. Department of Electrical Engineering, Delft University of Technology, Delft, the Netherlands
2. Billinton R, Allan RN (1996) Reliability evaluation of power systems, 2nd edn. Plenum Press, New York
3. Li W (2011) Probabilistic transmission system planning. Wiley Interscience - IEEE Press, Canada
4. Li W (2005) Risk assessment of power systems - models, methods, and applications. Wiley Interscience - IEEE Press, Canada
5. Tuinema BW (2017) Reliability of transmission networks: impact of EHV underground cables and interaction of onshore-offshore networks, PhD thesis. Department of Electrical Engineering, Electrical Sustainable Energy, Delft University of Technology, Delft, The Netherlands
6. Tuinema BW, Gibescu M, Kling WL (2009) Reliability evaluation of offshore wind energy networks and the dutch power system, MSc Thesis. Department of Electrical Engineering, Delft University of Technology, Delft, the Netherlands
7. Lemstrom B et al (2008) Effects of increasing wind power penetration on power flows in European grids. Tradewind
8. Wu L, Park J, Choi J, El-Kaib AA, Shahidehpour M, Billinton R (2009) Probabilistic reliability evaluation of power systems including wind turbine generators using a simplified multi-state model: a case study. In: IEEE power and energy society general meeting 2009 (PES'09)
9. Madras N (2002) Lectures on Monte Carlo methods. American Mathematical Society, Province
10. Yates RD, Goodman DJ (2005) Probability and stochastic processes. Wiley Inc, United States
11. Sigmazone.com (2011) Understanding binomial confidence intervals. http://www.sigmazone.com/binomialconfidenceinterval.htm

Part III
Applications

Chapter 6
Probabilistic Power Flow Analysis

Modern power systems are quickly evolving as many fundamental changes are currently taking place. On the one hand, there is an accelerated market transformation. For example, the European Union (EU) is moving fast to a single European electricity market, thereby increasing the volume of energy trade between countries. As a consequence, the cross-border transmission capacity is expected to increase until it reaches a reasonable target of 15% electricity interconnection by 2030 [1]. On the other hand, there is an unprecedented utilization of renewable energy and other low-carbon technologics, like the impressive expected 392 GW of offshore wind power to be installed in the North Sea, together with more *Demand Side Management* (DSM) and the new competitive market for *Distributed Generation* (DG) [2]. A common factor in the previous driving forces is the increasingly rapid number of uncertainties coming to play a critical factor in the planning and operation of the modern power system. This chapter is dedicated to present a complete overview of uncertainty modeling and the mechanism of modeling this in the steady-state permeance analysis of electrical power systems.

Learning Objectives

After reading this chapter, the student should be able to:

- describe the main sources of uncertainty in electrical power systems;
- describe and perform elemental modeling of uncertainty: generation/demand;
- describe the main methods used in probabilistic power flow;
- understand the application of non-sequential Monte Carlo simulation applied to probabilistic power flow analysis.

This chapter has been written by dr. Francisco M. Gonzalcz-Longatt (full professor, dept. of Electrical Engineering, Information Technology and Cybernetics, faculty of Technology, Natural Sciences and Maritime Sciences, Universitetet i Sørøst-Norge, Norway) and dr.ir. José L. Rueda Torres (associate professor, Intelligent Electrical Power Grids (IEPG), dept. of Electrical Sustainable Energy (ESE), faculty of Electrical Engineering, Mathematics and Computer Science (EEMCS), Delft University of Technology, the Netherlands).

© Springer Nature Switzerland AG 2020
B. W. Tuinema et al., *Probabilistic Reliability Analysis of Power Systems*,
https://doi.org/10.1007/978-3-030-43498-4_6

6.1 Uncertainties in Power Systems

6.1.1 Sources of Uncertainty

Uncertainty may be defined as the lack of knowledge regarding the real value of a parameter [3]. This concept frequently appears in power system modeling, as parameters of system models, system variable patterns and correlation between system components may never be known precisely. Unplanned outages, fault types, fault locations and wind power production are just a few of such uncertainties. The occurrence of unplanned outages by component failures and the impact on the reliability of power systems has already been discussed extensively in Chaps. 2–5 of this book. This chapter focuses on the uncertainty aspects of input parameters like the load and (renewable) generation in a power system. Parameter uncertainty in modeling has two primary sources: (i) randomness, when the variability of a specific phenomenon or all factors affecting the system are not known precisely and (ii) incompleteness, purely due to a lack of information regarding the parameter values.

Figure 6.1 shows a schematic representation of the parameter uncertainties, together with the analysis and modeling aspects involved. The idea of uncertainty modeling and simulation is being able to describe the uncertainty in the input parameters and to measure the impact of these on the output parameters. Uncertainty caused by randomness in the system inputs can be addressed by collecting data and performing *statistical analysis*. Then, the uncertainty in the input parameters is described using a probabilistic description, for example a *Probability Density Function* (PDF). However, if the data is incomplete or statistical data is not available (e.g. for future technologies under development), the situation becomes really complex. A possible solution is to make assumptions based on *human experts*, which might provide possibilities on parameter values, which is the case of cognitive uncertainties.

The sources of uncertainty in future power systems are located across the whole system: generation, network and demand. Uncertainties coming from the demand side are dominant in the traditional power system, while there is less proportion of randomness coming from classical dispatchable generators. Forecasting the electric consumption in the future power system will be more challenging than nowadays, as the increased time and spatial variation in load and its composition are expected to increase the uncertainty related to load models and parameters.

Fig. 6.1 Classification of uncertainties in power systems, including modeling and analysis aspects

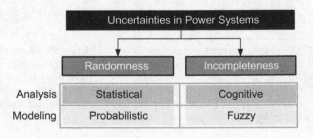

The accelerated penetration of non-fully dispatchable Renewable Energy Sources (RES), like solar and wind generation, injects a substantial amount of uncertainty at the generation side of the power system. The generation portfolio in future power systems will be characterized by extremely high uncertainty caused by the generation pattern as a consequence of size, type, location and output of the generation units. Also, the mixture of conventional generation, renewable and another sorts of power injection, such as electric mobility and balancing mechanism like energy storage and interconnectors, add uncertainty in future power systems.

Network uncertainties have been a problem in power systems since the very beginning of the commercial use of electricity. This source of uncertainty can come in very different ways: topology, parameters and settings. The topology uncertainties mainly come from the configuration uncertainty, which is evident in the case of contingencies. Power system faults are a very wide source of uncertainty in several aspects, like the type, location, duration, frequency, distribution and impedance of the fault. As already discussed in Sect. 2.2.4 of this book, component failures can be associated with various electrical faults. Other network uncertainties, like tap settings and temperature-dependent line rating, provide further uncertainty in the network's parameters and settings in network modeling. Moreover, the future energy system will be based on massive deployments of ICT, which can possibly also provide new sources of uncertainty to be considered, such as noise, time delays and loss of signals. Chapter 8 of this book will therefore discuss the reliability aspects of cyber-physical power systems.

6.1.2 Load Modeling: The Gaussian Mixture Model

The term *'load'* can have several meanings in power system engineering. Nevertheless, it is typically related to a very specific power consumption [4]:

- *device load*: a device connected to a power system that consumes a certain amount of power;
- *system load*: total (active and/or reactive) power consumed by all devices connected to a power system;
- *busbar load*: a portion of the system that is not explicitly represented in a system model, but rather treated as if it were a single power-consuming device connected to a bus in the system model;
- *generator or plant load*: the power output of a generator or a generating plant.

The loads are a very important element in the power system and are essential for power system analysis, planning and control. The phrase *'load modeling'* refers to the mathematical representation of the power consumption of a very specific load. Scientific literature has dedicated many publications to address the power dependence to two main electrical variables: voltage and frequency. Modeling of power system loads is a complex task, as it is influenced by many factors, including the dependence of the power consumption on other aspects, which relationships are frequently difficult to obtain deterministically. Load modeling requires much attention because the final model has a significant impact on almost every single regime of power system studies. Many classifications of load models can be found in literature [5–7].

A very generic and old classification of load models is based on the changes in the power demand over time, and two broad categories are defined: *static* and *dynamic models*. Static models express the active and reactive power at any instant of time as functions of the bus voltage magnitude and frequency at the same instant. These static models can be used to represent static loads (e.g. resistive loads) [5], and, as an approximation, dynamic loads (e.g. induction motors). The most common static model is the so-called ZIP model [4, 8], which represents the static load as a combination of constant power (P), constant current (I), and constant impedance (Z) elements [9]. Other static models include the exponential model [10], the frequency-dependent model [7], and the Electric Power Research Institute (EPRI) LOADSYN model [11]. Dynamic models express the active and reactive power as a function of voltage at past instants of time and, usually, including the present time instant [7]. Difference or differential equations can be used to represent such models. A summary of dynamic load models can be found in [5–8].

The objective of load modeling is to develop mathematical models to approximate load behavior. It would be desirable having the simplest model, but a balance between accuracy and simplicity of the models is often used. In general terms, load modeling consists of two main steps [5]: (i) selecting an appropriate load model structure and (ii) identifying the load model parameters using component or measurement-based approaches. The approach looks simple at first glance, but is extremely difficult in reality. Load modeling is a very challenging task, as it is influenced by a large number of diverse load components, the lack of precise load composition information and stochastic, time-varying and weather-dependent load behavior. The stochastic nature of the demand is a crucial issue of load modeling and it is becoming more and more challenging in future networks where the penetration level of distributed energy sources is high. Accurate models of the variability of the power system loads are essential for planning studies and operation.

There are several approaches to model the uncertainty of loads. However, the most common is to model the loads by a *Gaussian (or normal) distribution*. According to the Gaussian distribution, the PDF of the power demand (P_L) of a load can be expressed mathematically by:

$$f_{P_L}\left(P_L | P_L', \sigma_L^2\right) = \frac{1}{\sigma_L \sqrt{2\pi}} e^{-\frac{\left(P_L - P_L'\right)^2}{2\sigma_L^2}} \quad -\infty \leq P_L \leq \infty \text{ and } \sigma_L > 0 \qquad (6.1)$$

where P_L' is the mean value (or expectation) of the power demand and σ_L is the standard deviation. Stochastic simulation scenarios can be generated easily from the Gaussian distribution with appropriate assumptions about the distribution parameters (P_L' and σ_L).

Some scientific studies have demonstrated that a single Gaussian distribution is not always suitable for all loads since the statistical distribution of electric load variations may not strictly follow any standard probability distribution function [12]. For example, Fig. 5.8 in Chap. 5 of this book showed that the power demand of a typical domestic substation can better be modeled by a combination of two Gaussian distributions. A significant research effort has been devoted to probabilistic modeling of load variation by using different probability density functions (PDFs): normal (or Gaussian) [13, 14], log-normal [15], gamma [13], Gumbel, inverse-normal [13], beta [16, 17], exponential [18], Rayleigh, and Weibull [19]. An important conclusion that can be derived from literature survey is that there is not a unique or generalized technique to model the PDF of a load [20].

So far, Gaussian models of load profiles have been extensively used for various motives: (i) straightforwardness, as it can be described by only two parameters (i.e. the mean (μ) and the standard deviation (σ)), and (ii) the analysis of this PDF is the most developed and documented in literature. In recent times, publications in several areas such as finance, biometric, biology and, most recently, electrical engineering, have used the concept of a parametric PDF represented by a weighted sum of Gaussian component densities, which is called the *Gaussian Mixture Model* (GMM) [21]. The aforementioned domestic substation load model shown in Fig. 5.8 in Chap. 5 of this book therefore is an example of a GMM consisting of two normal distributions.

The Gaussian Mixture Model is a weighted sum of N_C component Gaussian densities as described by [22]:

$$p\left(\mathbf{x}|\lambda\right) = \sum_{i-1}^{N_C} w_i g\left(\mathbf{x}|\boldsymbol{\mu}_i, \boldsymbol{\Sigma}_i\right) \tag{6.2}$$

where $\mathbf{x}$ is a D-dimensional continuous-valued data vector (i.e. measurement or features), $w_i, i = 1, \ldots, N_C$, are the mixture weights, and $g\left(\mathbf{x}|\boldsymbol{\mu}_i, \boldsymbol{\Sigma}_i\right), i = 1, \ldots, N_C$, are the component Gaussian densities. Each component density is a D-variate Gaussian function of the form [22]:

$$g\left(\mathbf{x}|\boldsymbol{\mu}_i, \boldsymbol{\Sigma}_i\right) = \frac{1}{(2\pi)^{\frac{D}{2}} |\boldsymbol{\Sigma}_i|^{\frac{1}{2}}} e^{\left\{-\frac{1}{2}(x-\mu_i)' \boldsymbol{\Sigma}_i^2 (x-\mu_i)\right\}} \tag{6.3}$$

with mean vector $\boldsymbol{\mu}_i$ and covariance matrix $\boldsymbol{\Sigma}_i$. The mixture weights satisfy the constraint that the sum of all the weights must equal to one [22]:

$$\sum_{i=1}^{N_C} w_i = 1 \tag{6.4}$$

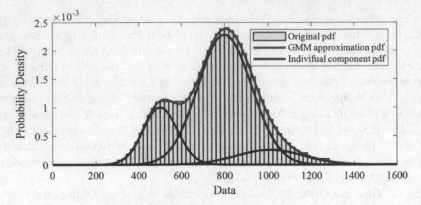

Fig. 6.2 Probabilistic distribution density for a hypothetical load: data created synthetically ($w =$ [0.7 0.1 0.2], $\mu =$ [800 1008 500], $\sigma^2 =$ [15000 24000 6400])

This condition is because a PDF must be non-negative and the integral of a PDF over the sample space of the random quantity it represents must be unity. A PDF for a hypothetical load consisting of a mixture of three components ($N_C = 3$) is presented in Fig. 6.2.

6.1.3 Load Modeling: Gaussian Mixture Model—Example

Singh et al. introduced the statistical modeling of the loads in distribution networks through a Gaussian Mixture Model [20], and the *Expectation-Maximization* (EM) algorithm was used to obtain the parameters of the mixture components. The GMM parameters are usually estimated via the EM algorithm, however, despite its conceptual simplicity, the EM algorithm may have difficulties in handling problems with high dimensionality [22]. The problem of obtaining various mixture components (e.g. weight, mean and variance) can be formulated as a problem of identification where optimization methods provide an efficient solution. The research paper titled *'Identification of Gaussian Mixture Model using Mean-Variance Mapping Optimization: Venezuelan Case'* [22], has proposed an application of the *Mean-Variance Mapping Optimization* (MVMO) algorithm to the identification of the parameters of GMMs representing the variability of power system loads.

The GMM example described in the section considers the Venezuelan power system. Venezuela's power system is an integrated vertical power company, called *Corporación Electrica Nacional* (Corpoelec), which covers most of the country [22]. The Paraguaná Peninsula transmission system is fed from a single circuit (230 kV) transmission line of San Isidro substation as part of the Venezuelan power pool. The average demand of the Paraguaná power system is around 280 MW, and import by the San Isidro tie line is approximately 200 MW. Figure 6.3 shows the configuration

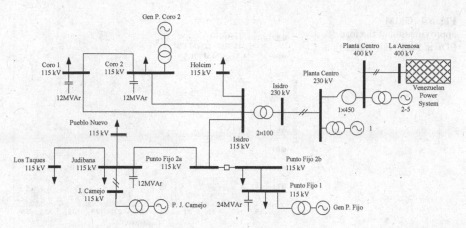

Fig. 6.3 Representative single-line diagram of the regional transmission system of the Paraguaná area in Venezuela

of the main components of the Paraguaná power system, including all transmission substations, transmissions lines, static reactive compensators and generators [23].

The most important loads of the Paraguaná power system are located at the Punto Fijo and Judibana substations. For this reason, these substations are selected as illustrative examples of the Gaussian Mixture Model representation. The time series from real measurements of the active power demand during two years (on an hourly basis) of those substations are used to create the GMM.

Figures 6.4 and 6.5 show the full component merging of the weighted mixture components, which was obtained by using the MVMO approach proposed in [22]. The discrete PDF of the real data is presented together with the GMM models created using MVMO and the EM method. The GMM model provides an excellent approximation of the statistical data collected at these two substations of the Paraguaná Transmission System. The GMM fit obtained through the EM algorithm is also included in the figures in order to validate the results obtained via MVMO. The parameters of each GMM component (obtained with the proposed approach) are summarized in Tables 6.1 and 6.2.

This estimation can be easily performed in MATLAB® using the *gmdistribution.fit* command. Note the closeness between the models identified using both MVMO and EM, which indeed highlights the accuracy that can be achieved with the proposed approach. Furthermore, it is worth to mention that the chi-square goodness-of-fit for the MVMO and EM cases was 11.35 and 22.12, respectively, for the Judibana estimate. Despite the small difference, it can be concluded that the MVMO estimate provides better estimation accuracy. These results suggest that any load PDF, irrespective of its distribution, can be estimated with a high degree of confidence using the proposed approach.

Fig. 6.4 GMM
approximation of the load
PDF at Punto Fijo Substation

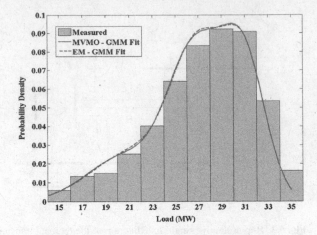

Fig. 6.5 GMM
approximation of the load
PDF at Judibana Substation

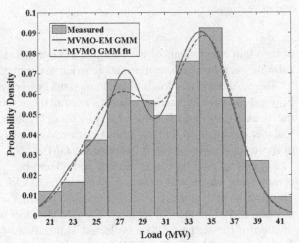

Table 6.1 Results of GMM approximation of the load PDF at Punto Fijo Substation [22]

Gaussian PDF (no.)	Weight, w (p.u.)	Mean, μ (MW)	Std, σ (MW)
1	0.1790	32.4068	9.2641
2	0.1956	27.6210	2.0460
3	0.4649	34.4864	6.8014
4	0.1605	24.1748	4.5416

Table 6.2 Results of GMM approximation of the load PDF at Judibana Substation [22]

Gaussian PDF (no.)	Weight, w (p.u.)	Mean, μ (MW)	Std, σ (MW)
1	0.1665	29.0589	4.8444
2	0.1936	20.6211	10.7520
3	0.1939	26.1117	4.5117
4	0.1922	26.8740	6.3281
5	0.2539	31.1581	3.1041

6.2 Power Flow Analysis

6.2.1 Deterministic Power Flow Analysis

Deterministic power flow analysis is used to analyze and assess the planning and operating of a power system on a daily routine [24]. It utilizes specific values for power generation and load demand in a selected network configuration in order to calculate system states and power flows [25]. The core of deterministic power flow analysis is solving a set of equations that defines the *power balance* of the power system (i.e. generation = demand + losses):

Power balance equations:
$$\begin{cases} \sum_{k=1}^{N_{\text{buses}}} P_{\text{gen}} - \left[\sum_{k=1}^{N_{\text{buses}}} P_{\text{load}} + P_{\text{losses}} \right] = 0 \\ \\ \sum_{k=1}^{N_{\text{buses}}} Q_{\text{gen}} - \left[\sum_{k=1}^{N_{\text{buses}}} Q_{\text{load}} + Q_{\text{losses}} \right] = 0 \end{cases} \tag{6.5}$$

The mathematical problem is typically determining a *static power flow*, as it is used to describe the power system performance at a particular condition where the system is operating in steady state, and algebraic variables are used to represent the network model. However, the equations are not linear, principally because of the power-voltage relationship. As a consequence, it is extremely hard to get analytical solutions of the deterministic power flow analysis and in fact, a numerical solution is the preferred option.

The most general form of the power flow equations, Eq. (6.5), is a set of *nonlinear algebraic equations* in steady state [26]:

$$\mathbf{g}(\mathbf{x}, \mathbf{u}) = \mathbf{0} \tag{6.6}$$

where **g** represents the set of nonlinear algebraic equations defining the power balance of system buses, **x** is the state vector containing the *state variables* (x_i), and **u** is a vector representing the set of *independent variables or inputs variables*. Then, the formulation of the power flow equations is reduced to the algebraic equation:

$$\left[\mathbf{g}^{P}\left(\mathbf{x}\right)\ \ \mathbf{g}^{Q}\left(\mathbf{x}\right)\right]^{T}=\mathbf{0} \tag{6.7}$$

Superscripts ^{P} and ^{Q} are used to split the active and reactive power, respectively. In general, the deterministic power flow problem elaborates in finding the zero of a set of nonlinear equations (i.e. the power-balance equations, Eq. (6.7)) emanating from an adequate initial guess ($\mathbf{x}_0$).

For the classical formulation of AC deterministic power flow (see also Appendix B of this book), the *inputs* or known quantities are the injected active powers (P_i) at all buses (where P and Q or P and V are known) excluding the slack bus, the injected reactive powers (Q_i) at all load buses (where P and Q are known), and the voltage magnitude at all generator buses (where P and V are *outputs* or known):

$$\begin{cases} P_i = g_i^P\left(\delta_1, \delta_2, \ldots, \delta_n, V_1, V_2, \ldots, V_n\right) \\ Q_i = g_i^Q\left(\delta_1, \delta_2, \ldots, \delta_n, V_1, V_2, \ldots, V_n\right) \end{cases} \tag{6.8}$$

where $i = 1, 2, \ldots, N_{\text{buses}}$ for N_{buses} power buses, and nonlinear voltage (V) and phase (δ) relationships. A complete explanation of classical AC power flow can be found in [27–29].

6.2.2   Probabilistic Power Flow Analysis

The randomness related to weather-dependent power production, like the fluctuating character of wind energy, causes the generation output of a wind power plant to be hardly controllable, neither continuous. Also, there is uncertainty in other components of the power system, such as the randomness of power demands. A critical drawback of deterministic power flow analysis is that it ignores any uncertainty related to the power system, such as component outages, network changes and load variations. Therefore, supporting the conclusions from the discussion of deterministic and probabilistic approaches in Sect. 1.3.2 of this book, the deterministic approach is not sufficient for the analysis of modern power systems penetrated by renewable energy sources like wind power. Consequently, the results from deterministic power flow analysis may lead to quite different, contradictory or even erroneous results. This may result in tremendous economic and technical impacts.

The need to include probabilistic uncertainties in power flow analysis motivated the rise of alternative methods to address the randomness related to some processes in the power flow analysis. In *probabilistic power flow analysis*, the power generation and the system configurations are both considered as discrete random variables, while the power demand is considered as a continuous random variable [30, 31]. Probabilistic power flow was first proposed in 1974 by Borkowska and Allan [32], to provide a full reflection of the influences on a range of different factors' random variations in power systems. Consequently, any uncertainties concerning the electric power demands of the loads, the power productions of a generation technology

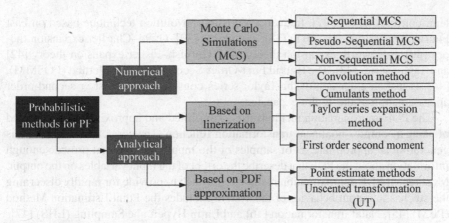

Fig. 6.6 Classification of probabilistic methods applied to PF

(especially wind power) and network parameters could be directly appeased through this method.

As illustrated in Fig. 6.6, probabilistic power flow analysis can be subdivided into two categories [22, 25]: (i) the *numerical approach* and (ii) the *analytical approach*. A combination of both approaches, i.e. a hybrid approach, is possible as well. The analytical approach analyzes the probabilistic power flow considering the inputs as purely mathematical expressions representing the uncertainty. Typical approaches use PDFs as a representation of the input variables, and the output results are obtained in the form of PDFs as well. The idea behind all the analytical approaches is to calculate the PDF of the output variables by using the arithmetic properties of the PDF and the rules defined by the power balance equations. Details of this approach can be found in [33, 34]. The numerical approach is to adopt a *Monte Carlo* (MC) method for the analysis. This substitutes a chosen number of values from the stochastic variables and parameters of the system model and performs a deterministic analysis for each value so that the same number of values are obtained in the results [25]. *Monte Carlo Simulation* (MCS) is used in [35, 36] to solve the probabilistic power flow problem, including wind farms penetration, by repeated simulations. The two main features of MC simulation are: (i) it provides considerably accurate results, but the computation time is consuming for large systems and (ii) it can be easily combined with pre-existent deterministic power flow programs to create an easy and fast implementation of probabilistic power flow. In this chapter, the approach selected is a probabilistic power flow based on MC simulation.

Literature Review of Probabilistic Power Flow

The analytical approaches as described above can be classified considering the approximation used in the solution: (i) *linearization* and (ii) *PDF approximation*. Several methods have been proposed in literature for analytical approaches to solving the probabilistic power flow using linearization, all of them looking into a drastic reduction in the computational burden of solving the probabilistic power flow problem: convolution [33, 37], frequency-domain convolution technique based on Fast Fourier Transformation (FFT) [38], cumulants [39], Gram–Charlier expansion theory [40], Edgeworth expansion theory [41], Cornish–Fisher expansion theory [42] and Taylor series [43]. The Hybrid First-Order Second-Moment Method (FOSMM), [44], is just an extension of the Taylor series considering the first or second-order approximation of the Taylor expansion.

The PDF approximation methods use a simplified and approximate PDF instead of using a complex nonlinear transformation function. The core of those methods is generating an appropriate set of samples of the input variables that provide enough information about the PDF that describe the effect of the input variables on the output. Most recent PDF approximation methods approach to provide for rapidly discerning the stochastic output behavior of the system includes the Point Estimation Method (PEM) [45], Nataf transformation [46] and Latin Hypercube Sampling (LHS) [47].

The most used and accurate numerical approach in probabilistic power flow analysis is the Monte Carlo (MC) approach. MC simulation (MCS) is a sampled-based method that is very commonly used because it can be used independently of the system characteristics, which makes it very useful in the highly nonlinear case of power systems. The MC simulation can be classified depending on the state sampling methods in three main groups: (i) *sequential MCS*, (ii) *non-sequential MCS* and (iii) *pseudo-sequential MCS*. Similar to the discussion in Sect. 5.3.1 of this book, the main difference between those methods is the way of preserving the time-dependent characteristic of the series used to sample the states of the power system. MC simulation has been effectively used to calculate the probabilistic power flow considering the variability of the power injections (e.g. wind power, river flows and load demand): sequential MCS is used in [48], non-sequential MCS in [35, 49] and pseudo-sequential MCS in [50].

Finally, uncertainties provided from incompleteness or lack of information regarding parameter values in generation and load can be considered in the power flow using the possibility approach. The possibility approach is based on the representation of power system uncertainties using membership functions and fuzzy arithmetic. Each membership function provides a degree of possibility of uncertainty. The main assumption in the possibility approach is that the result of the power flow can be represented by a possibly distribution or membership function instead of a PDF. The so-called *fuzzy load-flow model* has been successfully used in power systems [51–53]. The possibility of hybrid methods as hybrid possibilistic-probabilistic methods has been explored in other areas rather than probabilistic power flow [54].

Non-sequential MC Simulation Applied Probabilistic Power Flow

The probabilistic power flow analysis starts with the set of nonlinear algebraic equations in (6.6) and includes the power system randomness in the power flow formulation by adding this as a component of the input vector ($\mathbf{u}$). As a consequence, the input vector is redefined based on two components: one vector ($\boldsymbol{\lambda}$) containing the deterministic variables (λ_i) and a vector ($\mathbf{X}$) containing the *random variables* (X_i) to model power system uncertainties.

Fig. 6.7 Flowchart of MC
simulation applied to solve
probabilistic power flow

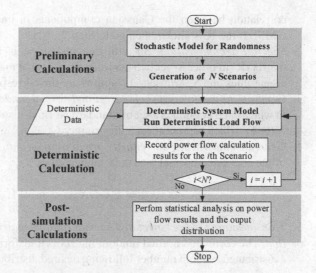

$$\mathbf{u} = [\lambda \ \mathbf{X}]^{\mathrm{T}} \qquad (6.9)$$

The output of the probabilistic power flow solution process consists of PDFs representing the state variables (x_i) and other dependent variables, e.g. branch elements active power flows (P_{jk}) or reactive power injection (Q_k) [55]. The MC simulation can be described in several ways, but all of these consider a systematic, iterative process [56]. In this chapter, the approach used for the MC simulation is an iterative process where the deterministic power flow is solved considering a set of input variables coming from non-sequentially generated random scenarios, called simulation scenarios. This approach is straightforward and can be implemented easily in already existing deterministic power flow software [57, 58]. Figure 6.7 shows the details of the process of solving the probabilistic power flow using non-sequential MC simulation used in this chapter.

Generation of Simulation Scenarios

Monte Carlo simulation requires to iteratively solve the deterministic power flow using scenarios where uncertainties are represented using pseudo-random numbers that are distributed according to a given probabilistic model (with or without correlation). The ith *simulation scenario* is a vector, $\mathbf{X}^i$, containing pseudo-random values of the n inputs uncertainties, $\mathbf{X}^i = \left[X_1^i, X_2^i, \ldots X_n^i \right]^{\mathrm{T}}$. There are various ways to generate multiple (i.e. more than two) sequences of correlated pseudo-random values. The two most common methods are: (i) a *Cholesky decomposition* of the correlation matrix ($\mathbf{R}$), (ii) *eigenvector decomposition* of the correlation matrix (also known as *spectral decomposition*). The latter approach is used in this chapter.

The procedure followed consists of five steps:

1. *Calculate the correlation matrix* ($\mathbf{R}$), *if not provided*: The matrix $\mathbf{R}$ of correlation coefficients (r_{ij}) is calculated from an uncertainties matrix $\underline{\mathbf{X}}$ whose rows are observations and whose columns are uncertainties (n). It is assumed a unique

correlation between the Gaussian components in uncertainties represented by GMM. If the **R** is known then go to Step 2.

2. *Spectral decomposition*. Spectral decomposition is applied to the correlation matrix (**R**) in order to get two matrices: **D**, diagonal matrix containing the $(n \times n)$ eigenvalues and **V**, whose (n) columns are the corresponding right eigenvectors.

3. *Generate the multivariate non-correlated normal random matrix*. $\mathbf{X}_{\text{ncorr}}$ is a matrix in which each column is a set of Standard Normal-distributed generated numbers, $N(0, 1)$, which represent each uncertainty (X_i).

4. *Generate multivariate correlated normal random matrix* ($\mathbf{X}_{\text{corr}}$). The mth column of the correlated normal matrix is calculated as:

$$X_{\text{corr},m} = V \sqrt{D} X_{\text{corr},m} \quad m = 1, \ldots, n \qquad (6.10)$$

5. *Transformation of the correlated normal random matrix into the desired distribution*. The correlated normal random matrix is transformed into a matrix (**X**) with a distributed random number following desired distributions for each uncertainty:

   a. *Normally Distributed Uncertainty*: Standardized normal distribution of each uncertainty is reverted to the appropriate mean and standard deviation $N(\mu, \sigma)$.

   b. *Non-Normally Distributed Uncertainty*: The transformation starts calculating the *Cumulative Distribution Function* (CDF) of each uncertainty and then the inverse probability integral transform or Smirnov transform is used to obtain the desired distribution.

In the case of uncertainties modeled using a GMM, the previous procedure can be applied. However, standardization might be reverted using a procedure related to 5a but considering $\underline{n}$-components and its weights.

6.3   Probabilistic Power Flow Example

This section illustrates the application of uncertainty modeling and calculation of the steady-state probabilistic power flow in future power systems. The single PDF modeling of randomness approach is used to represent each of the power injections, and 50,000 simulation scenarios are generated. MATLAB scripts have been created to represent the scenarios of each power injection, MATLAB R2019b has been used for this purpose. The test system is created using DIgSILENT® PowerFactory™ 2019 Service Pack 3, and a specialized script is created using DIgSILENT Programming Language (DPL) in order to perform the probabilistic power flow using MC simulation.

6.3.1 Test System Description

An intentionally modified version of the well-known 'IEEE 39-bus system' is used. The original system consists of 10 synchronous machines and 39 buses. The system is representative of the New England area in the North of the United States [59, 60]. The original system has been used to study small signal stability in a power system with a highly symmetrical structure. Today, the IEEE-39 tests system, as it is widely known, has been used in virtually all possible power system studies. The test system has been modeled using DIgSILENT® PowerFactory™. The configuration of the test system can be seen in Fig. 6.8.

The original test system has been customized to represent a hypothetical scenario of 40% (∼2438.84 MW) integration of new generation technologies, where the presence of uncertainties has been modeled using probabilistic data. New generation technologies, as well as Electrical Energy Storage Systems (EESS) and Plug-in Hybrid Electric Vehicles (PHEV), are included. A summary of the installed capacity of these technologies is given in Table 6.3.

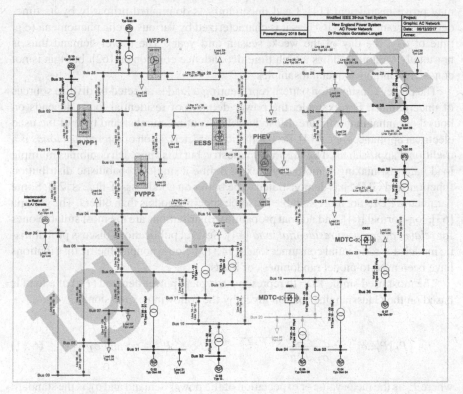

Fig. 6.8 Test system: modified version of IEEE 10-generator, 39-bus system, AC network, new technologies are highlighted in red color

Table 6.3 Details of the installed capacity of the active elements in the test system

Code	Type	Bus	Installed
WPP1	Onshore wind farm	25	615
WPP2	Offshore wind farm	40	615
PHEV	Power Hybrid Electric Vehicle	16	400
EESS	Energy Storage (Battery)	17	200
PVPP2	Photovoltaic Power Plant	2	800
PVPP1	Photovoltaic Power Plant	3	400

6.3.2 Probabilistic Model of Generic Power Demand (Loads)

Load randomness is the primary source of uncertainty in traditional electrical power systems. *Independence* and *normality* are two assumptions that offer an excellent modeling approach for the uncertainty in load demands. This approach is extensively used in the analytical formulation of probabilistic power flow, but this approach may prove insufficient [14]. Load uncertainty is dominated primarily by its time-dependent randomness [61], as it is characterized by various cyclic phenomena (e.g. time of day, the day of the week, season and year). The power demand thus is not totally random and has a high time-dependence component [62], but this is not considered in non-sequential samples.

The power consumption pattern representing a load is affected by diverse sources of uncertainties. For example, the power demand of residential loads depends on household inhabitant mixture, lifestyle or life routine, number and type of the used electrical appliances, etc. [14]. Probabilistic characterization of *aggregated loads* is a traditional approach used by the power industry, but this approach combines multiple load patterns, making it more complex to define a single probabilistic distribution function, and GMM has demonstrated to be better on goodness-of-fitness [22]. Some parametric PDFs are more suitable for some specific models than others. The gamma [63], log-normal [64], and normal probability distributions are the most suitable ones for *related aggregated residential load* [14]. Several publications discuss probabilistic models for residential consumer loads [14, 17]. Some nonparametric distributions have been used to model randomness of loads.

The most used single PDF to represents a typical power demand (P_L) of a load is based on the Gaussian distribution given by the following expression:

$$f_{P_L}\left(P_L|P_L', \sigma^2\right) = \frac{1}{\sigma_L\sqrt{2\pi}}e^{-\frac{(P_L - P_L')^2}{2\sigma_L^2}} \quad -\infty \le P_L' \le \infty \text{ and } \sigma_L > 0 \quad (6.11)$$

where P_L' is the mean value or expectation of the power demand and σ_L is the standard deviation. Stochastic simulation scenarios can be easily generated from the Gaussian

Fig. 6.9 Box-and-whisker
diagram and scatterplot of
power demand (P_L)

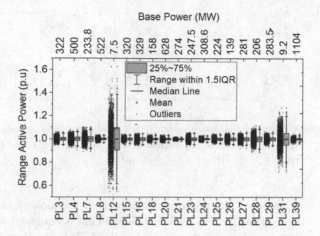

distribution with appropriate assumptions about the distribution parameters (P'_L and σ_L).

The randomness of the power demands (P_L) in the test system is modeled using a two-parameter (P'_L, σ_L) Gaussian (or normal) PDF. Numerical values of the non-sequentially generated random scenarios are shown in Fig 6.9.

Electrical Mobility: Plug-in Hybrid Electric Vehicles

Relatively new technologies like Plug-in Hybrid Electric Vehicles (PHEVs) are increasing their penetration in the power system. This technology is subject to different types of uncertainties, like charging periods, charging levels, battery start/end status and battery capacity. Gan Li et al. [73] established a probabilistic model describing the charging (non-controllable) demand of a PHEV highly dependent on binomial distribution theory. Therefore, a *binomial* PDF can be used in such cases where a more detailed and complicated model, as presented in [74], cannot be used.

The half-hourly power demand of a PHEV can be described based by two parameters: (i) *Start Charging Time* (*SCT*) and (ii) *State Of Charge* (*SOC*) at the beginning of the *SCT*. The energy consumption of a PHEV is proportional to the distance driven (D). Consequently, the randomness in the *SOC* may be expressed as:

$$SOC(D) = SOC_0 - D \qquad (6.12)$$

A reasonable assumption is that the PHEV is fully charged on a daily basis ($SOC_0 = 1.0$) [73]. This chapter considers a non-sequential sampling for the randomness in the distance driven (D), which follows a beta distribution function with parameters a and b.

$$f_{\text{PHEV}}(D|a,b) = \frac{\Gamma(a+b) D^{a-1} (c-D)^{b-1}}{\Gamma(a)\Gamma(b) c^{a+b-1}} \quad D \geq 0,\ a,\ b,\ c > 0 \qquad (6.13)$$

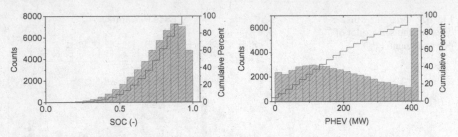

Fig. 6.10 Histogram and discrete CDF of SOC of PHEV and power demand ($a_{PHEV} = 1.47$, $b_{PHEV} = 5.09$), $SOC_{min} = 0.4$, $SOC_{max} 0.6$)

The most developed and used battery technology in PHEV is the lithium-ion (Li-ion) battery. The well-sold Nissan Leaf is equipped with that technology and its charging process is documented well [75, 76]. The charging power (P_{PHEV}) has two main regions: (i) a relatively constant demand during 4.5 h and then, (ii) the power quickly drops in the final 0.5 h.

Following the simple approximation of the two main aspects of the charging characteristic of the Li-ion battery, the charging power function is presented as:

$$P_{PHEV} = O \begin{cases} P_{max}, & SOC_{min} \leq SOC \leq SOC_{max} \\ \frac{P_{max}}{(1-SOC_{max})} (1 - SOC), & SOC_{max} \leq SOC \leq 1.0 \end{cases} \quad (6.14)$$

Figure 6.10 shows the histogram and discrete CDF of the SOC of the PHEVs in the test system. The values have been calculated using the probabilistic models presented above.

6.3.3 Probabilistic Model of Power Generation

The uncertainty in the power generation comes from both the output uncertainty due to the prime-mover/primary-resource randomness and the sudden disconnection of a power generator due to random failures. The power generation technologies used in the traditional power system are mainly dispatchable (e.g. nuclear, hydro and thermal), which means that the system operator can regulate the power output of the generators by controlling the supply of primary energy sources (or fuels). Uncertainties on dispatchable generation units are mainly related to unavailability. Single- or multiple-state power generation of a generation unit can be described by Probability Mass Functions (PMFs), as discussed in detail in literature [65].

For the considered test system, the randomness of the power injection coming from traditional dispatchable power plants has been included in four generators of the Test System, and it is simulated using discrete multiple-state PDF (see Fig. 6.11).

The integration of Renewable Energy Sources (RES), such as wind and solar power, introduces a substantial volume of uncertainty at the generation side of the

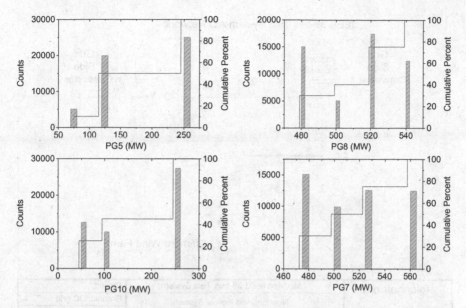

Fig. 6.11 Histogram and discrete CDF of power generation of classic power plants (PG5 (3-states, 254 MW), PG7 (4-states, 560 MW), PG8 (4-states, 540 MW), PG10 (250 MW, 3-states Power)

power system. The output from a RES-based power plant is not deterministic as a consequence of the uncertainty of the primary energy resource. Therefore, a probabilistic model must be used to describe the randomness of the production. The next paragraphs discuss the models of wind power plants and solar photovoltaic systems.

Wind Power Plants

The probabilistic model representing the power production of *Wind Power Plants* (WPPs) is well-documented in scientific literature [66]. The presented model uses a two-stage process: modeling the variability of the wind speed and then use the characteristics of the WPP to calculate the output power. The wind speed (v_w) provides for the randomness of the primary energy source of the WPP. The uncorrelated wind speed distribution, at a very specific location, can be represented as a two-parameter *Weibull distribution* [67]:

$$f_{v_w} = (v_w | a, b) = \frac{a v_w^{a-1}}{b^a} e^{-\left(\frac{v_w}{b}\right)^a} \quad v_w \geq 0, \ a > 0, \ b > 0 \tag{6.15}$$

where a is known as the Weibull scale parameter (in m/s) or slope, and it is proportional to the mean wind speed. b is the so-called Weibull shape parameter, which specifies the shape of the Weibull distribution.

The simplest probabilistic model of WPP production is an *aggregated model*, where a single equivalent generator characterizes the performance of the whole WPP. Such a model provides the total power production of the WPP as an integrated whole

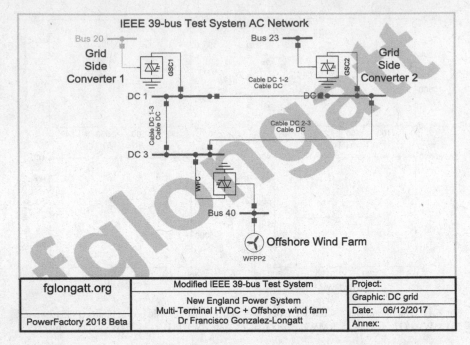

Fig. 6.12 Multi-terminal DC (MTDC): three-terminal VSC-HVDC transmission system used to integrate WPP2 into the test system

without distinguishing between each *Wind Turbine* (WT) in the WPP or including losses in the internal distribution system. Then, the probabilistic model for the output power of a wakeless wind farm is the sum of all wind turbine outputs obtained by combining the wind speed distribution with the wind turbine output curve [66, 67]. Based on the previous assumptions, the corresponding WPP output power production (P_{WPP}) is calculated using the power coefficient (C_p) of the WT and the number of them in the WPP array (N_{WT}):

$$P_{WPP}(v_w) = N_{WT} C_p(v_w) \qquad (6.16)$$

In this chapter, the customized test system includes two WPPs. Repower M6 6.15 MW wind turbine is used for illustrative purposes; each WPP consists of 200×6.15 MW. WPP1 is an onshore wind farm directly connected to bus 25. WPP2 is an offshore wind farm connected to a new AC bus with number 40. WPP2 is assumed to be located in a remote area and 3-terminal VSC-HVDC is used to transport the power to the test system at buses 20 and 23 (see Fig. 6.12).

In the test system, a two-parameter Weibull PDF is used to represent the wind speed (v_w) uncertainty. The wind speeds at each wind farms location are assumed slightly different regarding the shape and scale parameters of the PDFs and a weak

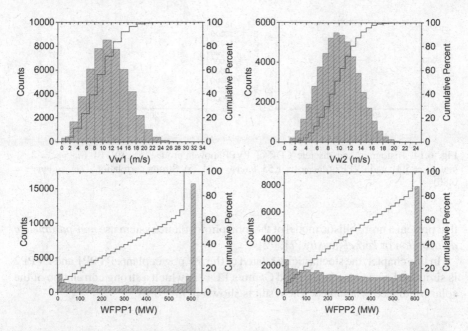

Fig. 6.13 Histogram and discrete CDF of wind speed (top) and power production (bottom). WPP1 ($a_{\text{WPP1}} = 12.826\,\text{m/s}$, $b_{\text{WPP1}} = 2.76$) and WPP2 ($a_{\text{WPP2}} = 11.123\,\text{m/s}$, $b_{\text{WPP2}} = 3.17$). Weak correlation: ($\rho_{\text{WPP1,WPP2}} = 0.25$)

correlation on the wind speed is assumed. Figure 6.13 shows the generation of the wind farms in the test system.

Photovoltaic Power Plants

The electrical generation pattern representing the power production in a *Photovoltaic* (PV) power plant is subject to different types of uncertainties, such as the irregular presence of clouds, solar irradiation and atmospheric temperature. Several publications present detailed methods to cope with the randomness uncertainties related to PV generation [68, 69]. The simplest probabilistic model to represent the uncertainty associated with *solar irradiance* is using a *Beta PDF*.

The output power (P_{PVPP}) of a Photovoltaic Power Plant (PVPP) is proportional to the number of panels and the solar irradiance. Then, the probabilistic model of the total power output of a PVPP is approximated by the beta PDF:

$$f_{P_{\text{PVPP}}}(p|a, b) = \frac{\Gamma(a + b)\, p^{a-1}\, (c - p)^{b-1}}{\Gamma(a)\, \Gamma(b)\, c^{a+b-1}} \quad 0 \leq p,\ c \text{ and } a\, b > 0 \qquad (6.17)$$

The ratio ($p = P_{\text{PVPP}}/P_{\text{PVPP,max}}$) between the total active power generated (P_{PVPP}) and the hypothetical power generated at maximum irradiance ($P_{\text{PVPP,max}}$) and the shape factors (a, b) of the *Gamma function* (Γ) define the probabilistic model of the PVPP. More details of this stochastic model can be found in [70]. More recent studies

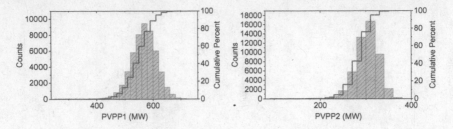

Fig. 6.14 Histogram and discrete CDF of PVPP power production. PVPP1 ($a_{PVPP1} = 2.33$, $b_{PVPP1} = 5.3$), and PVPP2 ($a_{PVPP2} = 1.55$, $b_{PVPP2} = 4.25$). Strong correlation: ($\rho_{PVPP1,PVPP2} = 0.80$)

that present a probabilistic model of the total photovoltaic system use *non-parametric distribution* or *kernel density* [71, 72].

In this chapter, the stochasticity related to the PV power plants PVPP1 and PVPP2 is simulated using a parameterized Gamma PDF, in which a strong correlation of the solar irradiance is assumed. The result is shown in Fig. 6.14.

6.3.4 Probabilistic Model of Balancing Technologies

Future energy systems will make use of several technologies not totally deployed or modestly used today, including several technologies to deliver counter-balance mechanisms to tackle the variability and intermittency of stochastic power generation. In this section, two balancing technologies are presented: *Electric Energy Storage Systems* (EESS) and *Multi-Terminal High-Voltage Direct Current* (MTDC) transmission systems.

Electric Energy Storage Systems
The electrical pattern representing the power production/consumption of an Electric Energy Storage System (EESS) depends on energy conversion and energy storage technology (e.g. mechanical energy storage, magnetic energy storage and chemical energy storage). There are several uncertainties related to the power production/consumption of an EESS, but the most important of all is the *energy charge state* (*EoC*) and its chronological information. An effort to overcome the chronological modeling for energy storage devices is presented in [77], and concludes that the *EoC* of a sustainable controlled EESS approaches a normally distributed function. However, the appropriate selection of the distribution parameters is open to discussion. The most used approach is to set a zero-power mean value of the power demand distribution and the standard deviation to the maximum power demand/production.

Figure 6.15 shows the histogram and discrete CDF of the *SOC* of an EESS. For the test system, the values have been calculated using the probabilistic models as described above.

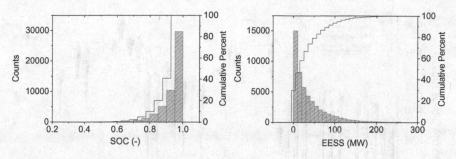

Fig. 6.15 Histogram and discrete CDF of the *SOC* and EESS power demand ($a_{EESS} = 0.70$, $b_{EESS} = 11.0$, $SOC_{min} = 0.0$, $SOC_{max} = 0.60$)

High Voltage Direct Current Links

High Voltage Direct Current (HVDC) transmission systems have been used for years in power systems in point-to-point connections using *Line-Commutated Converters* (LCC). However, technical development allows a progressive growth of HVDC based on *Voltage Source Converters* (VSC). VSC-HVDC systems are preferred for bulk power and integration of renewables because of their controllability, together with other features.

Power injection randomness in HVDC links depends on the configuration and operation mode of the VSC-HVDC system. A simple point-to-point HVDC link, designed for constant power flow between two operational areas, can be modeled by a *uniform probabilistic distribution* considering the system's unavailability.

A very specific PDF to model the power flow on a VSC-HVDC has been reported in [78]. It was created using data collected during a two-month period of operation of Murraylink (220 MW, $\pm$150 kV, VSC-HVDC), an embedded HVDC system within the National Electricity Market (NEM) of Australia. More complex probabilistic models are expected to rise from HVDC systems used to integrate offshore wind power coming from the North Sea for the multi-terminal configuration.

6.3.5 Simulation Results

For the considered test system, the numerical results of the probabilistic power flow simulation of 50,000 scenarios are post-processed to obtain the main statistical indicators of the state variables. The box-and-whisker diagrams and scatterplots of the bus voltages and power flows are presented in Figs. 6.16 and 6.17, respectively. Maximum, mean, minimum, median, range and interquartile range (IQR) are indicated in the graphs.

The statistical information of the voltages at two buses and the power flow of two branch elements are depicted for illustrative purposes in Figs. 6.18 and 6.19. The voltages and power flow distributions differ from the traditional Gaussian

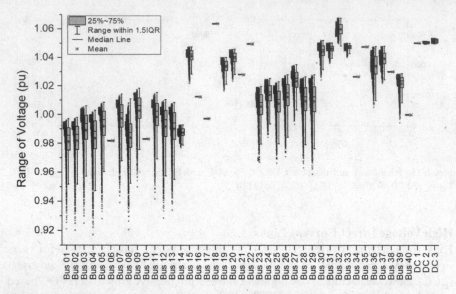

Fig. 6.16 Box-and-whisker diagram and scatterplot of bus voltages

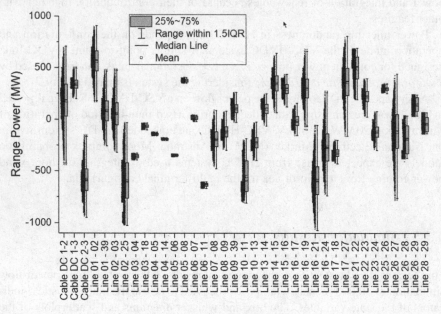

Fig. 6.17 Box-and-whisker diagram and scatterplot of transmission lines and DC cable power flow

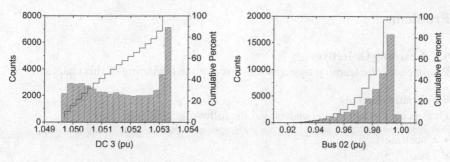

Fig. 6.18 Histogram and discrete CDF of bus voltage (pu) at DC3 (left) and Bus 02 (right)

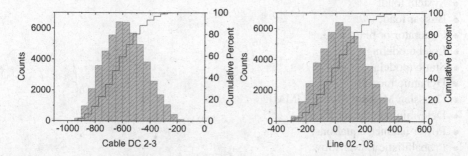

Fig. 6.19 Histogram and discrete CDF of power flow ($P_{kj} > 0$, from k to j in MW) at Cable DC 2–3 (left) and Line 02–03 (right)

distribution, which is caused by randomness of the power injections and the specific operation restrictions imposed by balancing technologies and electric mobility.

6.4 Conclusion

Modern power systems are systematically experiencing increasing levels of uncertainty, which are making operation and control of the system more complex. This chapter presented a comprehensive overview of uncertainty modeling and calculation of the steady-state power balance in the future power system. The methodology considered the uncertainties caused by the randomness on power injections to the power system and measures the impact of these in the steady-state power flow. A test system was used to illustrate the suitable use of the proposed methodology to obtain the probabilistic distribution of the system states, while numerical results illustrated the appropriateness of the proposed methodology.

Acknowledgements The authors would like to show their gratitude to DIgSILENT PowerFactory Germany for supporting this research and the related publications.

Problems

6.1 Learning Objectives
Please check the learning objectives as given at the beginning of this chapter.

6.2 Main Concepts
Please check whether you understand the following concepts:

- Uncertainty;
- Load;
- Device load;
- System load;
- Busbar load;
- Generator or plant load;
- Load modeling;
- Static model;
- Dynamic model;
- Gaussian Mixture Model (GMM);
- Deterministic Power Flow;
- Power balance equations;
- Probabilistic Power Flow;
- Monte Carlo (MC) simulation.

6.3 Uncertainties in Power Systems

- Explain the two primary sources of parameter uncertainty in power system modeling.
- Enumerate at least three of the most common models used to represent static loads.
- What is the most often used single PDF to represent power system demand? Please write down the equation.
- Write down the main equations used to represent a weighted sum of Gaussian probabilistic densities.

6.4 Probabilistic Power Flow

- Write down the equations used to represent the general problem of deterministic power flow (i.e. power balance equations in terms of active and reactive power).
- Enumerate the main categories of probabilistic power flow methods.

6.5 Non-sequential MCS applied PPF Problem

- Briefly explain the mechanism of applying MC simulation to probabilistic power flow.
- Explain how probabilistic power flow includes power system randomness in the analysis.
- Explain the steps used to generate non-sequential simulation scenarios used in MC simulation.

References

1. Gonzalez-Longatt F, Rueda JL (2018) Advanced smart grid functionalities based on powerfactory. Springer, UK
2. Gonzalez-Longatt F, Alhejaj S, Bonfiglio A, Procopio R, Rueda JL (2017) Inertial frequency response provided by battery energy storage systems: probabilistic assessment. In: 6th international conference on clean electrical power - renewable energy resources impact (ICCEP2017), pp 403–409
3. Rueda-Torres JL, Gonzalez-Longatt F (2017) Data mining and probabilistic power system security. Wiley, London
4. Task Force IEEE on Load Representation for Dynamic Performance (1993) Load representation for dynamic performance analysis (of Power Systems). IEEE Trans Power Syst 8(2):472–482
5. Arif A, Wang Z, Wang J, Mather B, Bashualdo H, Zhao D (2018) Load modeling - a review. IEEE Trans Smart Grids 9(6):5986–5999
6. IEEE Task Force on Load Representation for Dynamic Performance (1995) Bibliography on load models for power flow and dynamic performance simulation. IEEE Trans Power Syst 10(1):523–538
7. IEEE Task Force on Load Representation for Dynamic Performance (1995) Standard load models for power flow and dynamic performance simulation. IEEE Trans Power Syst 10(3):1302–1313
8. Concordia C, Ihara S (1982) Load representation in power system stability studies. IEEE Trans Power Appar Syst PAS-101(4):969–977
9. Kundur P, Balu NJ, Lauby MG (1994) Power system stability and control. McGraw-Hill, New York
10. Tseng KH, Kao WS, Lin JR (2003) Load model effects on distance relay settings. IEEE Trans Power Deliv 18(4):1140–1146
11. Price WW, Wirgau KA, Murdoch A, Mitsche JV, Vaahedi E, El'Kady M (1988) Load modeling for power flow and transient stability computer studies. IEEE Trans Power Syst 3(1):180–187
12. Limaye C, Whitmore DR (1984) Selected statistical methods for analysis of load research data - Final report. EPRI-EA-3467, USA
13. Carpaneto E, Chicco G (2006) Probability distributions of the aggregated residential load. In: 9th international conference on probabilistic methods applied to power systems (PMAPS06)
14. Carpaneto E, Chicco G (2008) Probabilistic characterisation of the aggregated residential load patterns. IET Gener Transm Distrib 2(3):373–382
15. Seppala A (1995) Statistical distribution of customer load profiles. In: Proceedings of the international conference on energy management and power delivery (EMPD95), vol 2, pp 696–701
16. Herman R, Kritzinger JJ (1993) The statistical description of grouped domestic electrical load currents. Elsevier Electr Power Syst Res 27(1):43–48
17. Heunis SW, Herman R (2002) A probabilistic model for residential consumer loads. IEEE Trans Power Syst 17(3):621–625
18. Lin HH, Chen KH, Wang RT (1993) A multivariant exponential shared-load model. IEEE Trans Reliab 42(1):165–171
19. Irwin GW, Monteith W, Beattie WC (1986) Statistical electricity demand modelling from consumer billing data. IEE Proc Gener Transm Distrib 133(6):328
20. Singh R, Pal BC, Jabr RA (2010) Statistical representation of distribution system loads using gaussian mixture model. IEEE Trans Power Syst 25(1):29–37
21. McLachlan G, Peel D (2000) Finite mixture models. Wiley, New York
22. Gonzalez-Longatt F, Rueda J, Erlich I, Villa W, Bogdanov D (2012) Mean variance mapping optimization for the identification of gaussian mixture model: test case. In: 6th IEEE international conference intelligent systems, pp 158–163
23. Gonzalez-Longatt FM (2013) Evaluation of reactive power compensations for the phase i of Paraguanã wind based on system voltages. In: Proceedings of the industrial electronics conference (IECON13)

24. Standards IEEE (1998) IEEE recommended practice for industrial and commercial power systems analysis (Brown Book). IEEE Std 399–1997, pp 1–488
25. Chen P, Chen Z, Bak-Jensen B (2008) Probabilistic load flow: a review. In: Proceedings of the 3rd international conference on electric utility deregulation and restructuring and power technologies, pp 1586–1591
26. Gonzalez-Longatt F, Roldan JM, Rueda JL, Charalambous CA (2012) Evaluation of power flow variability on the ParaguanÃą transmission system due to integration of the first venezuelan wind farm. In: Proceedings of the IEEE power energy society general meeting (PES-GM12), pp 1–8
27. Stagg GW, El-Abiad AH (1968) Computer methods in power system analysis. McGraw-Hill, New York
28. Brown HE (1985) Solution of large networks by matrix methods, 2nd edn. Wiley, New York
29. Arrillaga J, Arnold CP (1990) Computer analysis of power systems. Wiley, New York
30. Anders GJ (1990) Probability concepts in electric power systems. Wiley, New York
31. Mur-Amada J, Bayod-Rujula AA (2007) Wind power variability model part II - Probabilistic power flow. In: Proceedings of the 9th international conference on electrical power quality and utilisation (EPGU07), pp 1–6
32. Borkowska B (1974) Probabilistic load flow. IEEE Trans Power Appar Syst PAS-93(3):752–759
33. Allan RN, Borkowska B, Grigg CH (1974) Probabilistic analysis of power flows. Proc Inst Electr Eng 121(12):1551–1556
34. Leite da Silva AM, Ribeiro SMP, Arienti VL, Allan RN, Do Coutto Filho MB (1990) Probabilistic load flow techniques applied to power system expansion planning. IEEE Trans Power Syst 5(4):1047–1053
35. Jorgensen P, Christensen JS, Tande JO (1998) Probabilistic load flow calculation using Monte Carlo techniques for distribution network with wind turbines. In: Proceedings of the 8th international conference on harmonics and quality of power, vol 2, pp 1146–1151
36. Ramezani R et al (2011) Probabilistic load flow in distribution systems containing dispersed wind power generation. IEEE Trans Power Syst 16(2):1551–1556
37. Allan RN, Al-Shakarchi MRG (1976) Probabilistic a.c. Load Flow. Proc Inst Electr Eng 123(6):531–536
38. Allan RN, Leite da Silva AM, Burchett RC (1981) Evaluation methods and accuracy in probabilistic load flow solutions. IEEE Trans Power Appar Syst PAS-100(5):2539–2546
39. Schellenberg A, Rosehart W, Aguado J (2005) Cumulant-based probabilistic optimal power flow (P-OPF) with Gaussian and Gamma distributions. IEEE Trans Power Syst 20(2):773–781
40. Pei Z, Lee ST (2004) Probabilistic load flow computation using the method of combined cumulants and gram-charlier expansion. IEEE Trans Power Syst 19(1):676–682
41. Morrow D, Gan L (1993) Comparison of methods for building a capacity model in generation capacity adequacy studies. In: Proceedings of the IEEE WESCANEX93 - communications, computers and power in the modern environment, pp 143–149
42. Stuart A (2010) KendallâĂŹs advanced theory of statistics, vol 1. Distribution theory, 1st edn. Wiley, Chichester
43. Dhople SV, Dominguez-Garcia AD (2012) A parametric uncertainty analysis method for markov reliability and reward models. IEEE Trans Reliab 61(3):634–648
44. Madrigal M, Ponnambalam K, Quintana VH (1998) Probabilistic optimal power flow. In: Proceeding of the IEEE Canadian conference on electrical and computer engineering, vol 1, pp 385–388
45. Chun-Lien S (2005) Probabilistic load-flow computation using point estimate method. IEEE Trans Power Syst 20(4):1843–1851
46. Morales JM, Baringo L, Conejo AJ, Minguez R (2010) Probabilistic power flow with correlated wind sources. IET Gener Transm Distrib 4(5):641–651
47. Yan C, Jinyu W, Shijie C (2013) Probabilistic load flow method based on nataf transformation and latin hypercube sampling. IEEE Trans Sustain Energy 4(2):294–301

48. Lopes VS, Borges CLT (2015) Impact of the combined integration of wind generation and small hydropower plants on the system reliability. IEEE Trans Sustain Energy 6(3):1169–1177

49. VallÃl'e F, VersÃlle C, Lobry J, Moiny F (2013) Non-sequential Monte Carlo simulation tool in order to minimize Gaseous pollutants emissions in presence of fluctuating wind power. Elsevier Renew Energy 50(Supplement C):317–324

50. Manso LAF, Leite da Silva AM, Mello JCO (1999) Comparison of alternative methods for evaluating loss of load costs in generation and transmission systems. Elsevier Electr Power Syst Res 50(2):107–114

51. Wang Z, Alvarado FL (1991) Interval arithmetic in power flow analysis. In: Proceedings of the power industry computer application conference, pp 156–162

52. Das D, Ghosh S, Srinivas DK (1999) Fuzzy distribution load flow. Electr Mach Power Syst 27(11):1215–1226

53. Miranda V, Matos MA, Saraiva JT (1990) Fuzzy load flow - new algorithms incorporating uncertain generation and load representation. In: Proceedings of the 10th power system computation conference, Graz, Austria, pp 621–627

54. Soroudi A (2012) Possibilistic-scenario model for DG impact assessment on distribution networks in an uncertain environment. IEEE Trans Power Syst 27(3):1283–1293

55. Williams TJ, Crawford C (2010) Probabilistic power flow modeling: renewable energy and PEV grid interactions. In: Proceedings of the Canadian society for mechanical engineering forum (CSME FORUM 2010), Victoria, British Columbia, Canada

56. Aien M, Hajebrahimi A, Fotuhi-Firuzabad M (2016) A comprehensive review on uncertainty modeling techniques in power system studies. Elsevier Renew Sustain Energy Rev 57(Supplement C):1077–1089

57. Ramezani R, Ghavami SM (2011) Probabilistic load flow considering wind generation uncertainty. ETASR-Eng Technol Appl Sci Res 1(4):126–132

58. Gonzalez-Longatt F, Alhejaj S, Marano-Marcolini A, Rueda Torres JL (2018) Probabilistic load-flow analysis using DPL scripting language. Green energy and technology, no. 9783319505312, pp 93–124

59. Pai MA (1989) Energy function analysis for power system stability. Kluwer Academic Publishers, Boston

60. Athay T, Podmore R, Virmani S (1979) A practical method for the direct analysis of transient stability. IEEE Trans Power Appar Syst PAS 98(2):573–584

61. Abdullah MA, Agalgaonkar AP, Muttaqi KM (2013) Probabilistic load flow incorporating correlation between time-varying electricity demand and renewable power generation. Elsevier Renew Energy 55:532–543

62. Papaefthymiou G (2006) Integration of stochastic generation in power systems. University of Patras, Greece

63. Xia L, Alpcan T, Mareels I, Brazil M, De Hoog J, Thomas DA (2015) Modelling voltage-demand relationship on power distribution grid for distributed demand management. In: Proceedings of the 5th Australian control conference (AUCC15), pp 200–205

64. Ahmad A, Azmira I, Ahmad S, Nur Ilyana Anwar A (2012) Statistical distributions of load profiling data. In: Proceedings of the IEEE international power engineering and optimization conference, Melaka, Malaysia, pp 199–203

65. Billinton R, Allan RN (1996) Reliability evaluation of power systems, 2nd edn. Plenum Press, New York

66. Ackermann T (2012) Wind power in power systems, 2nd edn. Wiley, Chichester

67. Yuan Y, Zhou J, Ju P, Feuchtwang J (2011) Probabilistic load flow computation of a power system containing wind farms using the method of combined cumulants and gram-charlier expansion. IET Renew Power Gener 5(6):448–454

68. Ruiz-Rodriguez FJ, Hernandez JC, Jurado F (2012) Probabilistic load flow for radial distribution networks with photovoltaic generators. IET Renew Power Gener 6(2):110–121

69. Khallat MA, Rahman S (1986) A probabilistic approach to photovoltaic generator performance prediction. IEEE Trans Energy Convers EC-1(3):34–40

70. Karaki SH, Chedid RB, Ramadan R (1999) Probabilistic performance assessment of autonomous solar-wind energy conversion systems. IEEE Trans Energy Convers 14(3):766–772

71. Ren Z, Yan W, Zhao X, Zhao X, Yu J (2014) Probabilistic power flow studies incorporating correlations of PV generation for distribution networks. J Electr Eng Technol 9(2):461–470

72. Yan W, Ren Z, Zhao X, Yu J, Li Y, Hu X (2013) Probabilistic photovoltaic power modeling based on nonparametric Kernel density estimation. Autom Electr Power Syst 37(10):35–40

73. Sexauer JM, McBee KD, Bloch KA (2013) Applications of probability model to analyze the effects of electric vehicle chargers on distribution transformers. IEEE Trans Power Syst 28(2):847–854

74. Vlachogiannis JG (2009) Probabilistic constrained load flow considering integration of wind power generation and electric vehicles. IEEE Trans Power Syst 24(4):1808–1817

75. Dhameja S (2001) Electric vehicle battery systems. Newnes, Boston

76. Kejun Q, Chengke Z, Allan M, Yue Y (2011) Modeling of load demand due to EV battery charging in distribution systems. IEEE Trans Power Syst 26(2):802–810

77. Klockl B, Papaefthymiou G (2005) An effort to overcome the chronological modeling methods for energy storage devices. In: Proceedings of the international conference on future power systems, pp 6

78. Preece R, Milanovic JV (2015) Risk-based small-disturbance security assessment of power systems. IEEE Trans Power Deliv 30(2):590–598

Chapter 7
Extra-High-Voltage Underground Cables

Extra-high-voltage (EHV) underground cables are becoming a preferred solution for new connections in transmission networks. The visual impact of high-voltage towers on the landscape and health concerns related to electromagnetic fields are the main drivers for society to prefer underground cables (UGCs), especially in densely populated areas. In the Netherlands, EHV underground cables have been installed recently as part of the Randstad380 project [1]. The configuration, i.e. double circuits of 2x2635 MVA with 2 individual cables per circuit phase, and the fact that these UGCs are part of the heavily loaded 380 kV transmission network make this project rather unique.

Whereas there is much experience with UGCs at lower voltage levels, not much is known yet about the behavior of EHV cables in large transmission networks. Therefore, aspects like resonance, steady state, and transient behavior have been studied in the Randstad380 project [2, 3]. The reliability of EHV cable connections and the impact on the reliability of large transmission networks have been considered as well [4–10]. Regarding the reliability aspects, the focus of previous research was mainly on the components of cable circuits, while a comprehensive analysis of the reliability of UGC connections and the impact on large transmission networks was still missing. This chapter presents an overview of the main results of the research on UGC reliability performed within the scope of the Randstad380 project.

> **Learning Objectives**
>
> After reading this chapter, the student should be able to:
> - Calculate the reliability of an underground cable connection and compare it with the reliability of an overhead line connection;
> - Understand the impact of underground cables on large transmission networks;
> - Define measures to improve the reliability of an underground cable connection.

© Springer Nature Switzerland AG 2020
B. W. Tuinema et al., *Probabilistic Reliability Analysis of Power Systems*,
https://doi.org/10.1007/978-3-030-43498-4_7

7.1 Reliability of Overhead Line and Underground Cable Connections

7.1.1 Failure Statistics

As already discussed in Chap. 2 of this book, finding the failure frequency and repair time of EHV underground cables is challenging because of the scarcity of suitable failure statistics.[1] Whereas the failure frequency and repair time of overhead lines (OHLs) can be easily derived from failure statistics from databases like the one for the Dutch power system [11], other sources must be consulted for underground cables. Failure frequencies and repair times of UGC components derived from a survey among European TSOs have been reported in [12]. In this survey, high and low estimates of the failure frequencies were calculated, which will be further referred to as 'TSOs high' and 'TSOs low' in this chapter. Table 7.1 gives an overview of the failure frequencies and repair times used in this study. Based on failure statistics from the Dutch power system, a dependent failure factor of 0.1 can be assumed for dependent double-circuit failures [9], such that the failure frequencies of dependent double-circuit failures become $0.1\times$ the failure frequencies of independent single-circuit failures. Although the dependent failure factor could be different for OHLs and UGCs, the amount of available failure statistics is too limited to proof a difference. The repair time of UGCs is initially assumed to be 730 h (1 month), based on experience with EHV cable failures in the field, while 336 h (2 weeks) can be used for sensitivity analyses [9].

Table 7.1 Failure frequencies and repair times of OHLs and UGC components

Component	Failure frequency			Repair time		
	(High)[a]	(Low)[a]	(Unit)	(High)	(Low)	(Unit)
EHV OHL	0.220	0.220	/100cctkm·y [11]	8	8	h [11]
EHV cable	0.120[b]	0.079[b]	/100cctkm·y [12]	730[c]	336[c]	h
EHV joint	0.035[d]	0.016[d]	/100comp·y [12]	730[c]	336[c]	h
EHV termination	0.168[d]	0.092[d]	/100comp·y [12]	730[c]	336[c]	h

[a] (TSOs) high and (TSOs) low estimations of the failure frequencies were defined in [12]
[b] The failure frequency per circuit km is based on a circuit with only 1 individual cable per phase
[c] The repair time of the cable connection is assumed 730 h (1 month) or 336 h (2 weeks), based on experience with EHV underground cable failures in the field
[d] The failure frequency of joints and terminations is measured per individual component (i.e. per single phase)

---

[1]This section is based on the work published in [9]. ©2015 IEEE, reprinted with permission.

7.1.2 Reliability Calculations of Connections

The reliability of overhead line and underground cable connections can be calculated using the failure statistics of the components. For this calculation, the rules for series connections of components as described in Sect. 4.1 are applied. This is done for single-circuit as well as for double-circuit connections.

Single Circuits

For a single-circuit overhead line (as illustrated in Fig. 7.1, configuration a), the failure frequency and unavailability are:

$$f_{l1} = l_{total} f_{line} \ [/y] \tag{7.1}$$

$$U_{l1} = f_{l1} \frac{r_{line}}{8760} = l_{total} f_{line} \frac{r_{line}}{8760} \ [-] \tag{7.2}$$

where:

f_{l1} = failure frequency of a single OHL circuit [/y]
l_{total} = total connection length [km]
f_{line} = failure frequency of an OHL circuit [/cctkm·y]
U_{l1} = unavailability of a single OHL circuit [−]
r_{line} = repair time of an overhead line [h]

To calculate the reliability of a fully cabled circuit (as shown in Fig. 7.1, configuration b), all the components of the cable circuit (i.e. cable parts, joints, and terminations) must be included. Joints are installed between the cable parts, and terminations are present at both ends of the cable circuit. Under the assumption that $f \sim \lambda$, it can be assumed that $f_{connection} \sim \sum f_{components}$. Recalling Example 3.9 in Chap. 3, this assumption holds if the cable connection and the repair are both not too long. The failure frequency and unavailability of a cable circuit then become:

$$f_{c1} \approx n_{ic} l_{total} f_{cable} + 3 n_{ic} \left(N_{cpart} - 1 \right) f_{joint} + 3 \cdot 2 n_{ic} f_{term} \ [/y] \tag{7.3}$$

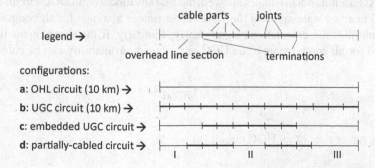

Fig. 7.1 Illustration of the 4 considered configurations (the total connection length is 10 km in all configurations; in configuration c and d, the total cable length is 5 km)

$$U_{c1} \approx n_{ic} l_{total} f_{cable} \frac{r_{cable}}{8760} + 3 n_{ic} \left(N_{cpart} - 1 \right) f_{joint} \frac{r_{joint}}{8760} + 3 \cdot 2 n_{ic} f_{term} \frac{r_{term}}{8760} \; [-]$$

$$(7.4)$$

where:

f_{c1} = failure frequency of a single UGC circuit [/y]
n_{ic} = number of individual cables per circuit phase [–]
l_{total} = total connection length [km]
f_{cable} = failure frequency of an underground cable
(single circuit, one cable per phase) [/cctkm·y]
N_{cpart} = number of cable parts [–]
f_{joint} = failure frequency of a single joint [/comp·y]
f_{term} = failure frequency of a single termination [/comp·y]
U_{c1} = unavailability of a single cable circuit [–]
r_{cable} = repair time of a cable (part) [h]
r_{joint} = repair time of a joint [h]
r_{term} = repair time of a termination [h]

If all cable parts have the same length (l_{cpart} [km]), the failure frequency and unavailability of a single UGC circuit can be calculated by:

$$f_{c1} \approx n_{ic} l_{total} f_{cable} + 3 n_{ic} \left\lceil \frac{l_{total}}{l_{cpart}} - 1 \right\rceil f_{joint} + 6 n_{ic} f_{term} \; [/y] \qquad (7.5)$$

$$U_{c1} \approx n_{ic} l_{total} f_{cable} \frac{r_{cable}}{8760} + 3 n_{ic} \left\lceil \frac{l_{total}}{l_{sec}} - 1 \right\rceil f_{joint} \frac{r_{joint}}{8760} + 6 n_{ic} f_{term} \frac{r_{term}}{8760} \; [-] \quad (7.6)$$

where:

$\lceil x \rceil$ = rounding x to the nearest integer towards infinity
l_{cpart} = length of a cable part [km]

When calculating the reliability of an UGC connection, it is often assumed that all UGC components have the same repair time. As the amount of available failure statistics is limited, it is not possible to accurately calculate the difference in repair times. As Example 4.8 in Chap. 4 showed, the Markov model of an UGC circuit can be reduced to a two-state model if the same repair time is assumed for all components. This simplifies the calculations significantly. Similarly, if the same repair time is assumed for all components of an UGC circuit, its unavailability can be calculated by:

$$U_{c1} = f_{c1} \frac{r_{cblcct}}{8760} \; [-] \qquad (7.7)$$

where:
r_{cblcct} = repair time of an UGC circuit [h]

Double Circuits

The occurrence of failures in double circuits can be derived from (7.1)–(7.7). The failure frequency and unavailability of independent double-circuit failures in double-circuit overhead lines are:

$$f_{12\text{ind}} = \binom{2}{1} U_{11} f_{11} = 2 \frac{f_{11} r_{\text{line}}}{8760} f_{11} = \frac{2}{8760} f_{11}^2 r_{\text{line}} \ [/\text{y}] \tag{7.8}$$

$$U_{12\text{ind}} = U_{11}^2 = \left(f_{11} \frac{r_{\text{line}}}{8760} \right)^2 = f_{11}^2 \frac{r_{\text{line}}^2}{8760^2} \ [-] \tag{7.9}$$

where:

$f_{12\text{ind}}$ = failure freq. of independent double circuit failures in overhead lines [/y]
$U_{12\text{ind}}$ = unavailability of independent double circuit failures in overhead lines [–]

The failure frequency and unavailability of independent double-circuit failures in UGC connections are:

$$f_{c2\text{ind}} = 2 U_{c1} f_{c1} \ [/\text{y}] \tag{7.10}$$

$$U_{c2\text{ind}} = U_{c1}^2 \ [-] \tag{7.11}$$

where:

$f_{c2\text{ind}}$ = fail. freq. of independent double circuit failures in UGC connections [/y]
$U_{c2\text{ind}}$ = unavail. of independent double circuit failures in UGC connections [–]

If it is assumed that all components of a UGC circuit have the same repair time (r_{cblcct}), these equations become:

$$f_{c2\text{ind}} = 2 U_{c1} f_{c1} = 2 \frac{f_{c1} r_{\text{cblcct}}}{8760} f_{c1} = \frac{2}{8760} f_{c1}^2 r_{\text{cblcct}} \ [/\text{y}] \tag{7.12}$$

$$U_{c2\text{ind}} = U_{c1}^2 = \left(f_{c1} \frac{r_{\text{cblcct}}}{8760} \right)^2 = f_{c1}^2 \frac{r_{\text{cblcct}}^2}{8760^2} \ [-] \tag{7.13}$$

To calculate the failure frequency and unavailability of dependent double-circuit failures, a dependent failure factor (c_{cc}) can be defined. The dependent failure factor is the ratio between the failure frequency of dependent double-circuit failures and the single-circuit failure frequency. The failure frequency and unavailability of dependent double-circuit failures then become:

$$f_{12\text{dep}} = c_{cc} f_{11} \ [/\text{y}] \tag{7.14}$$

$$U_{12\text{dep}} = f_{12\text{dep}} \frac{r_{\text{line}}}{8760} = c_{cc} f_{11} \frac{r_{\text{line}}}{8760} \ [-] \tag{7.15}$$

$$f_{c2\text{dep}} = c_{cc} f_{c1} \ [/\text{y}] \tag{7.16}$$

$$U_{c2\text{dep}} = f_{c2\text{dep}} \frac{r_{\text{cblcct}}}{8760} = c_{cc} f_{c1} \frac{r_{\text{cblcct}}}{8760} \ [-] \tag{7.17}$$

where:

f_{12dep} = fail. freq. of dependent double circuit failures in OHL connections [/y]
c_{cc} = dependent failure factor [–]
U_{12dep} = unavailability of dependent double circuit failures in OHL connections [–]
f_{c2dep} = fail. freq. of dependent double circuit failures in UGC connections [–]
U_{c2dep} = unavailability of dependent double circuit failures in UGC connections [–]

7.1.3 Failures in the Randstad380 Zuid Cable Connection

Using (7.1)–(7.17), the reliability of the 11-km-long Randstad380 Zuid UGC connection can now be analyzed. In this configuration, 2 individual cables are applied
per circuit phase to facilitate transport capacity ($n_{ic} = 2$). Furthermore, the length of
the cable parts is about 900 m ($l_{cpart} = 0.9$). As indicated in Fig. 7.1 (configuration b),
joints are installed at 12 locations, while terminations are located at both ends of the
connection. The high estimation of the failure frequencies from Table 7.1 is used
together with a repair time of 730 h. It is supposed that the dependent failure factor
is 0.1 for both underground cables and overhead lines.

First, the reliability of a single circuit is studied. Table 7.2 shows how the individual
components contribute to the total reliability of a single circuit. As can be seen from
the table, the cable circuit will fail about once every 14 years. The (annualized)
unavailability is 52 h/y then. All the components of the cable circuit contribute to the
total failure frequency and unavailability. As comparison, the total failure frequency
of an 11-km-long overhead line circuit is calculated. This overhead line circuit will
fail only once every 42 years, such that the cable circuit fails about three times as
often as the overhead line circuit. This is mainly caused by the two individual cables
per circuit phase and the additional joints and terminations. The unavailability shows
a large difference between the overhead line and the underground cable circuit, which
is mainly caused by the difference in repair times.

Table 7.2 Failure frequency and (annualized) unavailability of one circuit of the 11 km Rand-
stad380 Zuid cable connection using the TSOs high estimate of the failure frequency

Component	Amount	Fail. freq. [a]	f_{c1} [b] [/y]	U_{c1}^* [h/y]
Cable	11 km	0.00120 (×2)	0.026	19.3
Joints	72	0.00035	0.025	18.4
Terminations	12	0.00168	0.020	14.7
Total			0.071	52.4
Line	*11 km*	*0.00220*	*0.024*	*0.2*

[a] As the failure frequency of the cable is measured per circuit kilometer with a single cable per
circuit phase, this value is doubled for an UGC configuration with two cables per circuit phase
[b] f_{c1} is the total failure frequency of one circuit; U_{c1}^* is the annualized unavailability in hours/year;
the repair time of the cable connection is assumed 730 h

Table 7.3 Occurrence of failures in double-circuit connections

Failure type	Failure freq. [/y]		Probability* [h/y]	
	Line	Cable	Line	Cable
Single-circuit failures	0.048	0.142	0.4	104.8
Independent double cct	1.1e-6	8.3e-4	4.3e-6	0.31
Dependent double circuit	2.4e-3	7.1e-3	0.02	5.2

Table 7.3 shows the occurrence of failures in double circuits. The failure frequency and (annualized) probability of single-circuit failures are double the value as shown in Table 7.2, because there are two single circuits in a double circuit. Because of the long repair time of UGC connections, the unavailability of a single circuit is larger and it becomes more likely that the second circuit (independently) fails during the repair of the first circuit. Table 7.3 shows that independent double-circuit failures occur considerably more often in the cable connection than in the overhead line connection (8.3e-4 vs. 1.1e-6 /y). Also, the probability of having an independent double-circuit failure differs considerably (0.31 vs. 4.3e-6 h/y).

For dependent double-circuit failures, the table shows that these occur about three times as often in underground cable connections as in overhead lines (7.1e-3 vs. 2.4e-3 /y). This is similar to the difference found for single-circuit failures. Also, the difference in probability is comparable (5.2 vs. 0.02 h/y). This is because the same dependent failure factor is assumed for underground cables and overhead lines. It is interesting to compare the failure frequencies of dependent and independent double-circuit failures. Whereas this difference is several orders of magnitude for overhead lines (2.4e-3 vs. 1.1e-6 /y), this difference becomes smaller for underground cables (7.1e-3 vs. 8.3e-4 /y). Within UGC connections, independent double-circuit failures are more frequent because of the relatively long repair time and it becomes more likely that the second circuit fails independently during the repair of the first circuit.

This effect is studied in more detail for increasing connection lengths. Figure 7.2 shows the mean time between failures ($MTBF = 1/f$) and probability (annualized to h/y) of double-circuit failures in cable connections. As can be seen, from about 110 km, independent double-circuit failures will occur more often than dependent double-circuit failures. This can be explained by (7.12) and (7.16). These equations show that the failure frequency of dependent double-circuit failures increases linearly with the single-circuit failure frequency (and the connection length), whereas the failure frequency of independent double-circuit failures increases quadratically with the single-circuit failure frequency (and the connection length). The right graph shows that independent failures then contribute significantly to the total probability of having a double-circuit failure. For overhead lines, double-circuit failures are almost always dependent failures.

This can have consequences for the planning and operation of large transmission networks. In general, in transmission networks, it is assumed that single-circuit failures are relatively likely to occur. Dependent double-circuit failures occur less often,

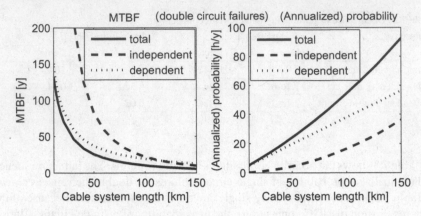

Fig. 7.2 Mean time between double-circuit failures and (annualized) probability of double-circuit failures in underground cable connections

and independent double-circuit failures are rare. If more and longer cable connections are installed in the network, independent double-circuit failures will happen more frequently. Possible independent double-circuit failures that include a cable circuit should therefore be studied in more detail during system design and observed closely during system operation.

7.1.4 Solutions to Improve the Reliability

There are several possible solutions to improve the reliability of a cable connection. Reducing the repair time or failure frequency of the components, partially cabling and alternative cable configurations are the studied solutions in this section.

Reducing the Failure Frequencies and Repair Time
In the previous calculations, the reliability of underground cable connections was calculated. Whereas the failure frequency of a cable circuit does not differ significantly from the failure frequency of an overhead line circuit, the repair time of a cable connection is about hundred times as long. Probably, the best way to increase the reliability of transmission networks with underground cables is to reduce the repair time of cable connections. In [12], the repair process of underground cables was analyzed, based on experience with UGC failures in the field and expert's opinion. Figure 7.3 shows the activities that contribute to the total repair time. As can be seen, the actual repair of the cable connection takes 1–2 weeks. However, activities like cleaning the site, getting spare/new parts, and arranging technical people can cause a much longer repair time.

If the repair process can be optimized more, the total repair time can be reduced significantly. This is illustrated in Fig. 7.4. Some repair activities can be planned

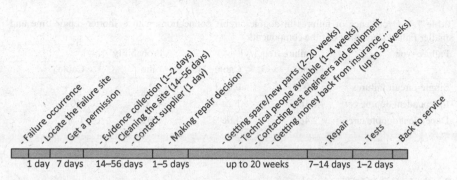

Fig. 7.3 Scheme of the repair time of cable connections (based on [12])

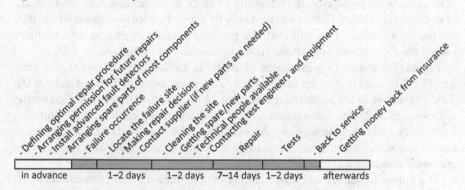

Fig. 7.4 Improved scheme of the repair time of cable connections

before a failure occurs. For example, permission to perform future repairs can be arranged beforehand. Also, spare parts of most components could be available for the case a failure occurs. Furthermore, with advanced fault detection, it becomes easier to localize the fault within limited time. Other activities like getting money back from the insurance can be managed after the cable connection is repaired. In this way, the repair time could be reduced to about 2 weeks (336 h), although exceptions are still possible.

Another way to increase the reliability of UGC connections is to reduce the failure frequency of the components [13]. Advanced installation and testing techniques are possibilities to reduce the failure frequencies.

The effects of smaller failure frequencies and a shorter repair time can be studied by using the equations from the previous section with the low failure frequencies from Table 7.1 and a repair time of 336 h (2 weeks). The results are shown in Table 7.4. As can be seen, the failure frequency and probability of failures within cable double circuits have reduced compared to Table 7.3. The difference between independent and dependent double-circuit failures in cable connections has increased, but there is still a large difference between overhead lines and underground cables.

Table 7.4 Occurrence of failures in double-circuit connections with a shorter repair time and smaller failure frequencies of the components

Failure type	Failure freq. [/y]		Probability* [h/y]	
	Line	Cable	Line	Cable
Single-circuit failures	0.048	0.080	0.4	58.4
Independent double cct	1.1e-6	1.2e-4	4.3e-6	0.02
Dependent double circuit	2.4e-3	4.0e-3	0.02	1.3

Partially Cabled Connections

Another solution to improve the reliability of an UGC connection is to limit the use of underground cables. Cables will then only be applied at those locations where they are the most desired. This will lead to a partially cabled connection. In a partially cabled connection, one or multiple cable sections can be embedded.

Figure 7.1 shows two examples of partially cabled connections. Configuration Fig. 7.1c is an embedded cable circuit. This is the situation in the Randstad380 Zuid connection in the Netherlands, where a 11 km cable section is installed within a connection of 22 km. Configuration Fig. 7.1d is a partially cabled connection with multiple embedded cable sections. An example is the Randstad380 Noord connection in the Netherlands, where three cable sections of 2, 3, and 3 km, respectively, are installed within a connection of 45 km.

The failure frequency and unavailability of a partially cabled connection can be calculated by combining (7.1)–(7.7):

$$f_{pc1} \approx n_{ic}l_{cable}f_{cable} + 3n_{ic}N_{csec}\left\lceil \frac{\frac{l_{cable}}{N_{csec}}}{l_{cpart}} - 1 \right\rceil f_{joint} + 6n_{ic}N_{csec}f_{term} + l_{line}f_{line} \ [/y]$$

$$(7.18)$$

$$U_{pc1} \approx \left(n_{ic}l_{cable}f_{cable} + 3n_{ic}N_{csec}\left\lceil \frac{\frac{l_{cable}}{N_{csec}}}{l_{cpart}} - 1 \right\rceil f_{joint} + 6n_{ic}N_{csec}f_{term}\right)\frac{r_{cblcct}}{8760}$$

$$+ l_{line}f_{line}\frac{r_{line}}{8760} \ [/y]$$

$$(7.19)$$

where:
$\lceil x \rceil$ = rounding x to the nearest integer towards infinity
f_{pc1} = failure frequency of a partially-cabled circuit [/y]
l_{cable} = total cable circuit length [km]
N_{csec} = number of embedded cable sections [–]
l_{line} = total overhead line length [–]
U_{pc1} = unavailability of a partially-cabled circuit [–]

In these equations, it is assumed that all components of the underground cable connection have the same repair time (r_{cblcct}). Furthermore, the number of joints is based on the total cable circuit length, the number of embedded cable sections, and the average cable part length. Depending on the length of the embedded cable sections, the number of joints might be somewhat different in reality.

The failure frequency and unavailability of double-circuit failures then become:

$$f_{pc2ind} = 2f_{pc1}U_{pc1} \ [/y] \tag{7.20}$$

$$U_{pc2ind} = U_{pc1}^2 \ [-] \tag{7.21}$$

$$f_{pc2dep} = c_{cc}f_{pc1} \ [/y] \tag{7.22}$$

$$U_{pc2dep} = c_{cc}U_{pc1} \ [-] \tag{7.23}$$

where:

f_{pc2ind} = failure frequency of independent double circuit
            failures in partially-cabled connections [/y]
U_{pc2ind} = unavailability of independent double circuit
            failures in partially-cabled connections [−]
f_{pc2dep} = failure frequency of dependent double circuit
            failures in partially-cabled connections [/y]
U_{pc2dep} = unavailability of dependent double circuit
            failures in partially-cabled connections [−]

Using Eqs. (7.1)–(7.7) and (7.18)–(7.19), the occurrence of single-circuit failures is calculated and shown in Table 7.5. It is assumed that the total connection length is 10 km, of which underground cable is 5 km. As can be seen, for the embedded cable circuit (configuration Fig. 7.1c), less joints are required. This leads to a smaller failure frequency and unavailability compared to configuration d.

It is also possible to divide the underground cables over multiple cable sections at different locations. For example in configuration Fig. 7.1d, there are two cable sections of 2.5 km each, within a connection of 10 km. As can be seen from Table 7.5, the main difference between configurations Fig. 7.1c and d is that in the partially cabled connection, additional terminations are required at 2 locations while one set of joints can be omitted. Compared to configuration Fig. 7.1c, both the failure frequency and unavailability have increased because of the additional terminations.

Table 7.5 Reliability comparison of partially cabled circuits

Configuration	Joints	Terminations	Failure frequency	Unavailability*
Figure 7.1a	–	–	0.0220 /y	0.176 h/y
Figure 7.1b	12 locations	2 locations	0.0694 /y	50.6 h/y
Figure 7.1c	5 locations	2 locations	0.0537 /y	31.2 h/y
Figure 7.1d	4 locations	4 locations	0.0717 /y	44.4 h/y

It is now interesting to compare the partially cabled circuit (configuration Fig. 7.1d) with the fully cabled circuit (configuration Fig. 7.1b). Table 7.5 shows that although the unavailability of the partially cabled circuit is smaller than the unavailability of the fully cabled circuit (44.4 vs. 50.6 h/y), the failure frequency of the partially cabled circuit is larger than the failure frequency of the fully cabled circuit (0.0717 vs. 0.0694 /y). It seems that for some cases, a fully cabled circuit performs better than a partially cabled circuit. If this is studied in more detail, there can be two cases:

1. A line section is located at the end of a connection (sections I and III in Fig. 7.1d)
2. A line section is located between two cable sections (section II in Fig. 7.1d)

If a line section at the end of a connection is replaced with cables, cables and joints are added to the connection while the amount of terminations remains the same. The unavailability of the connection will increase. If a line section in between two cable sections is replaced with cables, the difference is:

- -2 sets of terminations;
- $-l$ overhead line;
- $+2l$ underground cable;
- $+l/l_{\text{cp}} + 1$ sets of joints.

When this is compared using the parameters of Table 7.1 (high estimation), it is found that the unavailability of a 3.6 km overhead line section is equal to the unavailability of a 3.6 km cable section. Below 3.6 km, the unavailability of a line section is larger, and above 3.6 km, the unavailability of a cable section is larger. It seems therefore better to avoid overhead line sections shorter than 3.6 km.

Alternative cable configurations
Another option to improve the reliability of a cable connection is to use an alternative configuration. For example, the use of additional disconnectors, circuit breakers, or spare cables can be beneficial for the reliability.

In Fig. 7.5, some alternative configurations for single circuits are shown. Configuration Fig. 7.5o, a double-circuit overhead line, is the reference configuration. Configuration Fig. 7.5a is the standard Randstad380 cable configuration (with two

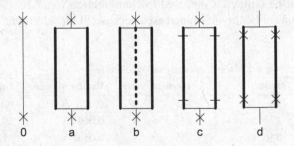

Fig. 7.5 Configurations of UGC single circuits (0 = single line circuit, a = single cable circuit (circuit switching), b = a with additional spare cable, c = single cable circuit with disconnectors (circuit switching), d = single cable circuit with circuit breakers (cable switching))

Table 7.6 Reliability of the alternative single-circuit configurations

	0	a	b	c	d
$P_{50\%}^*$ [h/y]	–	–	–	50.4	52.1
$P_{0\%}^*$ [h/y]	0.2	52.4	1.8	1.7	0.1
$f_{50\%}^*$ [/y]	–	–	–	0.071	0.071
$f_{0\%}^*$ [/y]	0.024	0.071	0.071	0.071	2.1e-4
$T_{50\%}^*$ [y]	–	–	–	14	14
$T_{0\%}^*$ [y]	41	14	14	14	4.7e3

$P_{50\%}^* / P_{0\%}^*$ = annualized probability of 50%/0% transmission capacity; $f_{50\%}/f_{0\%}$ = frequency of 50%/0% capacity states; $T_{50\%}/T_{0\%}$ = mean time between 50%/0% capacity states; $T = 1/f$

individual cables per circuit phase). In case of a cable failure, one circuit is disconnected by the protection system. In configuration Fig. 7.5b, an additional spare cable is installed that can be used after a cable failure. A switching time of 24 h is assumed. In configuration Fig. 7.5c, additional disconnectors are installed. After a failure, the damaged cable can be isolated and half a circuit can be put into operation again. In configuration Fig. 7.5d, additional circuit breakers are installed. After a cable failure, the damaged cable can be isolated directly and only half a circuit is disconnected.

Underground cable configurations can be effectively modeled by Markov models. For example, the Markov model as shown in Fig. 7.6 can be drawn for UGC configuration Fig. 7.5c. Using Markov models, the results as shown in Table 7.6 were obtained. The results from Table 7.2 are included in this table as well. As can be seen, if an additional spare cable is used (configuration Fig. 7.5b), the mean time between failure remains the same. This is because after a failure, the whole circuit is switched off first. But by using the spare cable, the circuit can be put into operation sooner and the total unavailability of the circuit is reduced. This unavailability is still higher than the unavailability of the line circuit.

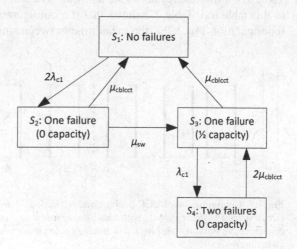

Fig. 7.6 Example of Markov model for the UGC circuit configuration as shown in Fig. 7.5c

If additional disconnectors are used (configuration Fig. 7.5c), the mean time between failures remain the same as well. It is however possible to continue operation at half capacity within a limited switching time. Consequently, the probability of the zero-capacity state is reduced. This half-capacity state is maintained until the damaged cable is repaired. With additional circuit breakers (configuration Fig. 7.5d), the cable circuit is switched per cable (half a circuit). The results in Table 7.6 show that the old values of the zero-capacity state have now become the new values of the half-capacity state. The probability of having zero transmission capacity has reduced significantly.

For system operation, the different configurations will have different consequences. In configuration Fig. 7.5b, it is important that the spare cable can be manually connected in a short time. A good design of the cable circuit and of the repair plan helps to reduce the repair time. In practice, the use of a spare cable causes some problems. In the calculation, it is assumed that the spare cable can replace any other cable of the circuit. In reality, this may be impossible because of the location of the cable and cross-bonding of the cables. Moreover, this spare cable must be tested regularly to guarantee that it can be used whenever needed. A last constraint is the additional costs that are probably larger than the benefits of the higher reliability of the cable circuit.

The configurations with the half-capacity states, configurations Fig. 7.5c and d, have consequences for system operation as well. The main constraint is the possibility of the half-capacity state. For example, when a cable circuit is part of a longer double-circuit overhead line circuit, a half-capacity state may cause problems. In case of a single failure, it may be necessary to take measures to reduce the power flow through the remaining circuit. If this is possible, the reliability of the cable circuit can be improved significantly with the investment of additional disconnectors or circuit breakers (and protection systems). Because of the half-capacity state, configuration Fig. 7.5d may perform even better than a single-circuit overhead line.

A similar study can be performed for double circuits, as illustrated in Fig. 7.7. Table 7.7 shows the results of the analysis. The results from Table 7.3 are included in this table too. Table 7.7 shows that if a configuration with a spare cable is used (configuration Fig. 7.7b), the mean time between single failures ($T_{50\%}$) remain the

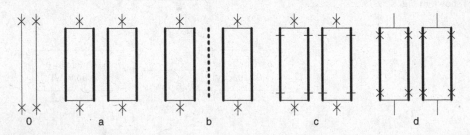

Fig. 7.7 Configurations of UGC double circuit (0 = double overhead line circuit, a = double cable circuit (circuit switching), b = a with additional spare cable, c = double cable circuit with additional disconnectors (circuit switching), d = double cable circuit with additional circuit breakers (cable switching))

Table 7.7 Reliability of the alternative double-circuit configurations

	0	a	b	c	d
$P^*_{75\%}$ [h/y]	–	–	–	103.5	103.5
$P^*_{50\%}$ [h/y]	0.39	103.5	3.87	50.4	0.46
$P^*_{25\%}$ [h/y]	–	–	–	2.1e-2	9.3e-4
$P^*_{0\%}$ [h/y]	4.3e-6	0.31	3.1e-5	1.7	6.9e-7
$f^*_{75\%}$ [/y]	–	–	–	0.14	0.14
$f^*_{50\%}$ [/y]	4.8e-2	0.14	0.14	0.14	1.3e-3
$f^*_{25\%}$ [/y]	–	–	–	8.3e-4	3.8e-6
$f^*_{0\%}$ [/y]	1.1e-6	8.3e-4	3.2e-5	1.3e-6	3.8e-9
$T^*_{75\%}$ [y]	–	–	–	7.1	7.1
$T^*_{50\%}$ [y]	20.7	7.1	7.1	7.1	786
$T^*_{25\%}$ [y]	–	–	–	1.2e3	2.6e5
$T^*_{0\%}$ [y]	9.3e5	1.2e3	3.1e4	7.9e5	2.6e8

$P^*_{75\%}/P^*_{50\%}/P^*_{25\%}/P^*_{0\%}$ = annualized probability of 75%/50%/25%/0% transmission capacity; $f_{75\%}/f_{50\%}/f_{25\%}/f_{0\%}$ = frequency of 75%/50%/25%/0% capacity states; $T_{75\%}/T_{50\%}/T_{25\%}/T_{0\%}$ = mean time between 75%/50%/25%/0% capacity states; $T = 1/f$; for independent failures only

same while the probability of single failures ($P^*_{50\%}$) becomes much smaller. Consequently, the MTBF of independent double failures ($T_{0\%}$) increases and the probability of independent double failures ($P^*_{0\%}$) is reduced.

If additional disconnectors are used (configuration Fig. 7.7c), the MTBF of the 50% capacity state remains the same ($T_{50\%}$). This is because after a failure, first one circuit is switched off. But because it is possible to continue operation at 75% within a limited switching time, the probability of the 50% capacity state is reduced ($P^*_{50\%}$). The old probability has now become the probability of the 75% capacity state ($P^*_{75\%}$). The MTBF of independent double-circuit failures can be increased, and the probability ($T_{0\%}$, $P^*_{0\%}$) can be reduced significantly in this way. With additional circuit breakers (configuration Fig. 7.7d), the MTBF of the 50% capacity state ($T_{50\%}$) is increased because after a failure, only half a circuit is switched off. The probabilities of having 50%, 25% or 0% capacity ($P^*_{50\%}$, $P^*_{25\%}$, $P^*_{0\%}$) become even smaller.

Similar to the single circuits, the use of a spare cable can cause practical problems in reality. The other configurations may be better alternatives. Whereas the use of additional circuit breakers (and protection systems) leads to higher overall costs, the use of additional disconnectors is a more efficient way to increase the reliability of the cable connection. It must however be studied whether the 75% capacity state is acceptable in the actual transmission network. In practice, a 75% capacity may be acceptable as cable circuits can be overloaded to a certain level and for a certain period. Dynamic loading could therefore be a solution to this.

7.2 Reliability Analysis of the Dutch Transmission Network

7.2.1 Approach of the Study

In this section, the reliability of the EHV transmission network in the Netherlands is studied. This analysis is based on state enumeration (as explained in Sect. 5.1), of which the specific algorithm is shown in Fig. 7.8. Contingencies are defined in contingency lists and include independent failures, dependent double-circuit failures, and dependent triple-circuit failures. Failure combinations up to 2 failure events are included. For each contingency and for each hour of the generation/load year scenario, a DC load flow is calculated. If an overload exists in the network, remedial actions are applied. First, national generator redispatch and, then, cross-border redispatch are performed. Load curtailment is performed as a last resort.

For this study, the EHV transmission network of the Netherlands as illustrated in Fig. 7.9 is used. The impact of EHV UGCs on the reliability of this network is analyzed in two steps. First, the reliability of a small part of the transmission network is studied, i.e. the region around Maasvlakte (MVL) substation, as indicated by the dashed ellipse in Fig. 7.9. In the second step, the reliability of the complete EHV transmission network is analyzed. For the load, export, wind, and conventional generation, a load flow scenario for 2020 is used. In this 'scenario2020,' values are given for every hour for each EHV substation.

7.2.2 Reliability of the Maasvlakte Region Network

In this study, the impact of UGCs at various locations within a transmission network is studied. For this purpose, the network connecting the large-scale generation center Maasvlakte (MVL) is considered. Figure 7.10 shows the configuration of this part of the transmission network. MVL substation is connected to the main 380 kV ring by two connection routes: MVL-WTL-WTR-BWK (Maasvlakte–Westerlee–Wateringen–Bleiswijk) and MVL-SMH-CST-KIJ (Maasvlakte–Simonshaven–Crayestein–Krimpen a/d IJssel). The first route is the new Randstad380 south connection, with a long UGC of 11 km between WTR and BWK and a shorter UGC of 2 km between MVL and WTL to cross a river. The connection MVL-SMH-CST-KIJ is an OHL connection. This situation will be referred to as 'base case.'

As shown in Fig. 7.10, about 2.5 GW of conventional generation is located at MVL substation. Another 0.5 GW generation is located at SMH. Moreover, the 1 GW submarine cable to the UK is connected to MVL. In the future, MVL will be one of the connection points of large-scale offshore wind energy. In scenario2020, 1 GW of offshore wind capacity is connected to MVL. The ranges of the busloads and the line loads to the main 380 kV ring are shown in the figure as well. The busloads of the substations are relatively small, such that the main load flow is from MVL to the main 380 kV ring. The studied network includes substation SMH. As SMH is a

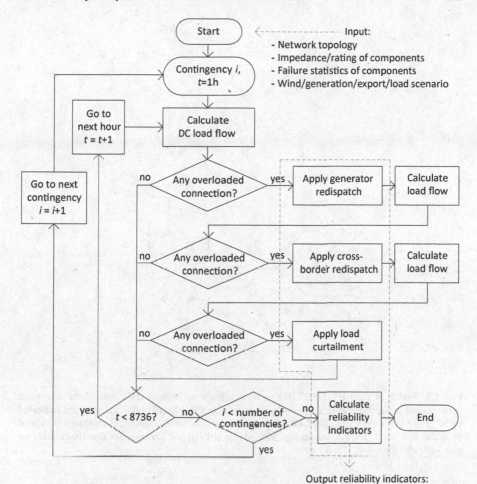

Fig. 7.8 State enumeration algorithm applied in this study (based on [7]). ©2016 IEEE, reprinted with permission

tapping station, a failure in a circuit connected to SMH will lead to the disconnection of the complete MVL-SMH-CST circuit. Therefore, MVL-SMH-CST is regarded as one (double circuit) connection in this study. For simplicity, SMH is not considered separately, but the load and generation at SMH are added to the load and generation at MVL.

It is interesting to study the impact of UGCs at various locations within this network. First, the situation in which the complete network consists of OHLs is studied. Then, certain OHL connections within the network are replaced with UGCs, until the complete network consists of UGCs. For each situation, the reliability of

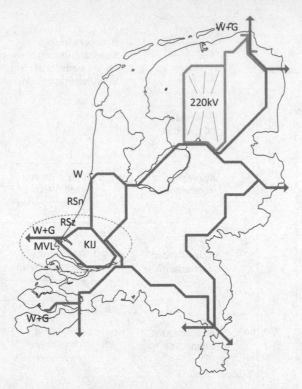

Fig. 7.9 Sketch of the EHV (380/220 kV) transmission network of the Netherlands. The main conventional generation centers and connection points of large-scale offshore wind are indicated by 'G' and 'W', respectively. The UGCs belonging to Randstad380 (north and south) are indicated by 'RSn' and 'RSz'. The two substations Maasvlakte (MVL) and Krimpen aan den IJssel (KIJ) are also indicated (based on [14, 15])

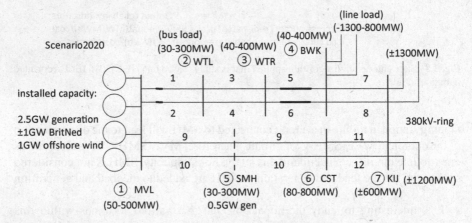

Fig. 7.10 Transmission network connecting MVL substation to the main 380 kV ring. In the base case, an 11 km double-circuit UGC is located between WTR and BWK and another 2 km between MVL and WTL to cross a river. These UGCs are indicated with bold lines

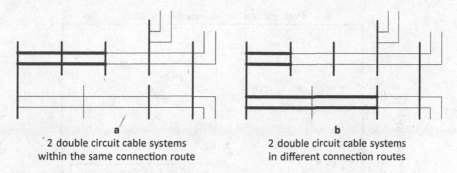

a
**2 double circuit cable systems
within the same connection route**

b
**2 double circuit cable systems
in different connection routes**

Fig. 7.11 Examples of 2 cable systems within the same and within different connection routes. As SMH380 is excluded from the study, the connection MVL-CST-KIJ is regarded as one double circuit

the network is calculated. Three more situations are considered: the base case (as shown in Fig. 7.10), the situation in which a double-circuit UGC of 1 km is installed within connection 1–2 ('1–2(1 km)'), and the situation in which 2 short double-circuit UGCs of 1 km each are installed within connections 1–2 and 9–10 ('2×1 km'). For multiple UGCs, a distinction between two situations can be made as illustrated in Fig. 7.11. In the first case, two UGCs are located within the same connection route between MVL and KIJ. In the second case, two UGCs are installed within different connection routes. This distinction will be important for the conclusions of this study.

First, the probability of an islanded substation within the network is calculated. A substation is islanded if it is not connected to the rest of the network anymore . Any substation within the network can become islanded. The probability of an islanded substation is a characteristic of the network topology and is independent of the load flow in the network. It can be calculated directly from the probabilities of the failure combinations that lead to an islanded substation.

Figure 7.12 shows the probability of an islanded substation. It can be seen that for increasing total UGC length, the probability of an islanded substation increases. Moreover, a distinction can be made between 0, 1, and 2 or more UGCs in the network. The graph also shows a difference between situations with and without UGCs in connection 9–10. If connection 9–10 is cabled, this contributes much to the total UGC length, while terminations (which contribute relatively much to the UGC unreliability) are only installed at two locations. Also, the probability of failure combinations (e.g. two dependent double-circuit failures) is smaller for one long cable than for multiple shorter UGCs. Consequently, it is still better to replace the complete connection 9–10 with UGCs than to install two 1 km double-circuit UGCs in connections 1–2 and 9–10 (2×1 km vs. 9–10). A substation is more likely to become isolated when it is connected by UGCs only. Figure 7.12 does not show any difference between multiple UGCs within the same and within different connection routes.

The network reliability is now analyzed with the algorithm of Fig. 7.8. It is found that load curtailment is never applied in this study. Because MVL is a large-scale

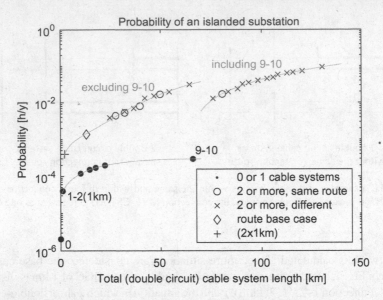

Fig. 7.12 Probability (in h/y) of an islanded substation within the MVL380 network

generation center, there are enough possibilities for generation redispatch. Load curtailment can still be the result of an islanded substation, but this has already been considered before. As the probability of load curtailment does not show any difference, the probability of redispatch is used instead. As discussed in Chap. 1, both load curtailment and redispatch can be a risk for a TSO.

The probability of redispatch with 1 GW installed offshore wind capacity at MVL is shown in Fig. 7.13. In general, the probability increases for a larger total UGC length. Several groups can be distinguished within the figure, depending on which connections include UGCs. The probability of redispatch is the largest for multiple UGCs including connections 1–2 and 9–10. As the main load flow within the network is from MVL to KIJ, the connections on the left in Fig. 7.10 are loaded slightly more than the connections on the right. Consequently, the connections on the left are more likely to become overloaded and will contribute more to the probability of redispatch. It can be concluded that the loading of an UGC connection influences the network reliability considerably. There is no clear difference between multiple UGCs within the same and within other connection routes in this case.

In the future, MVL will be one of the connection points of offshore wind energy. In scenario2020, 1 GW offshore wind capacity is connected to MVL. It is of interest to study the impact of more offshore wind. If the installed capacity of the offshore wind is increased to 3 GW, by scaling the wind generation in scenario2020, the maximum generation minus the load at MVL becomes about 5.2 GW. This is the maximum capacity of the network under $N-1$ redundancy. In this study, it is assumed that the additional wind power is absorbed by the main 380 kV ring at KIJ.

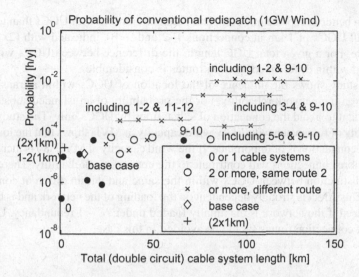

Fig. 7.13 Probability of redispatch within the MVL region network with 1 GW installed offshore wind capacity at MVL (the probability is annualized to h/y)

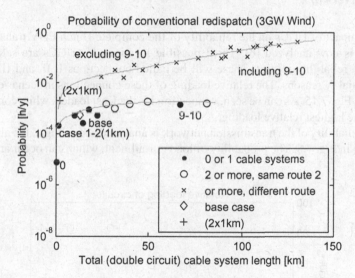

Fig. 7.14 Probability of redispatch within the MVL region network with 3 GW (right) installed offshore wind capacity at MVL (the probability is annualized to h/y)

The probability of redispatch for 3 GW installed wind capacity at MVL is shown in Fig. 7.14. As can be seen, the probability of redispatch has increased compared to the figure on the left. There is now a clear distinction between multiple UGCs within the same connection route and within different connection routes. According to the figure, multiple UGCs within the same connection route seem to act as one UGC.

It is even better to replace one complete connection route with UGCs than to install two small UGCs of 1 km in connections 1–2 and 9–10, indicated with (2x1km) in the figure. For a given total UGC length, the difference between UGCs within the same and within different connection routes is considerable.

This study shows the influence of the location of UGCs within a transmission network on the network reliability. For the probability of an islanded substation, it is important to avoid the connection of substations by UGCs only. The study on the probability of redispatch with 1 GW wind capacity at MVL shows that the loading of an UGC connection is of importance. If the wind capacity at MVL380 is increased to 3 GW, more connections will contribute to the network (un)reliability. There is then a clear distinction between UGCs within the same and within different connection routes. This effect is strongly dependent on the loading of the network and is the most pronounced if the network is maximally loaded under $N - 1$ redundancy. UGCs in different connection routes should be avoided in this case.

7.2.3 Underground Cables in the Dutch Transmission Network

The impact of UGCs on the reliability of the complete Dutch EHV transmission network is now analyzed [6]. Three possible locations for UGCs are selected to study the reliability impact. These will be called connections I, II, and III due to confidentiality reasons. The relative loading of these connections in scenario2020 is shown in Fig. 7.15. As can be seen, connection I is the least loaded, while connection III has the highest relative loading.

The reliability of the transmission network is analyzed according to the algorithm as shown in Fig. 7.8. The probability of load curtailment, which can occur anywhere

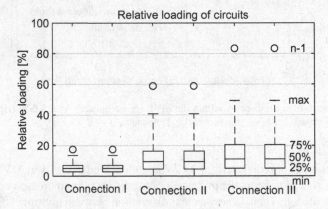

Fig. 7.15 Relative loading of the circuits of connections I, II, and III [7]. ©2016 IEEE, reprinted with permission

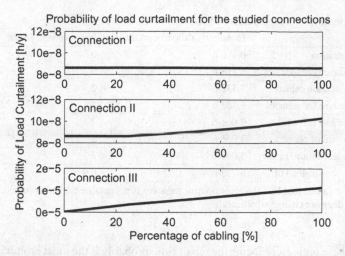

Fig. 7.16 Probability of load curtailment for underground cables in the studied connections [7]. ©2016 IEEE, reprinted with permission

within the network, is chosen here to represent the reliability. Within the studied connections, the total length of the UGCs is varied between 0% and 100% of the total connection lengths. As can be seen in Fig. 7.16, the probability of load curtailment does not change if UGCs are applied in connection I, whereas this probability increases if UGCs are installed in connection II. The probability of load curtailment increases considerably if UGCs are located in connection III. This difference is mainly caused by the different loadings of the three connections.

Several factors can influence the reliability of UGCs. The first factor is the application of series impedance compensation. UGCs have a smaller impedance than OHLs. Series impedance can be applied to better divide the load flow over the network. When studying this, it is found that the increase in probability of load curtailment of connection II is completely caused by the impedance difference. In contrast, series compensation of connection III does not give any improvement of the probability of load curtailment. It is therefore worth to study the effect of series impedance compensation for future UGC projects.

The sensitivity to various input parameters if 50% of connection III is cabled is shown in Table 7.8. As can be seen, series impedance compensation does not have any effect in this case. The network reliability could be increased by cabling only 25% of connection III. Other options are to reduce the failure frequencies or the repair time. If only half the transmission capacity is sufficient, the reliability can be improved by using only 1 cable per circuit phase. If 2 cable sections instead of 1 are installed in connection III, the reliability decreases by about 10%. The other way around holds as well: If 2 cable sections are installed, the reliability can be improved by about 10% by using only 1 cable section. Table 7.8 also shows that if the dependent failure factor for UGCs is 0.05 instead of 0.1, the probability of load curtailment reduces as well.

Table 7.8 Sensitivity to input parameters: connection III 50% cabled

Scenario	Probability of load curtailment	Improvement factor
Standard situation	6.11e-6	0
With impedance compensation	6.11e-6	0
25% of connection cabled	3.53e-6	2
TSO low failure frequencies	3.38e-6	2
Repair time 336h (2 weeks)	3.06e-6	2
1 cable per circuit phase[a]	3.06e-6	2
0.05 dependent failure factor	3.06e-6	2
2 cable sections instead of 1	6.79e-6	0.9

[a]Here, it is assumed that half the transmission capacity is acceptable while the impedance of the connection does not change significantly

From these options, reducing the repair time probably is the most promising, as the repair time can be reduced considerably by using a more optimized repair scheme. Although reducing the failure frequencies gives about the same improvement, it is more challenging to reduce these by 50%. Limiting the amount of UGC by a shorter cable length, less cable sections, and less cables per phase results in a better reliability, but these must be in line with the other requirements for a specific UGC project. Another option to reduce the 'repair time' of an UGC connection is to use a configuration with additional disconnectors. In this way, half a circuit can be put into operation after a cable fault within a limited switching time. The reduced capacity of the UGC connection should then be acceptable during system operation.

7.3 Conclusion

In this chapter, the reliability of EHV underground cables was studied. Starting with failure statistics, the reliability of underground cable (UGC) connections was analyzed and compared with the reliability of overhead line connections. Then, the impact on the reliability of large transmission networks was studied.

As 380 kV underground cables are a relatively new technology, the amount of available failure statistics is limited. Failure statistics from a survey among European TSOs were discussed at the beginning of this chapter. Comparing the reliability of EHV underground cables with the reliability of EHV overhead lines, it is found that UGCs are less reliable. The failure frequency of an UGC connection is about three times the failure frequency of an OHL connection. This is caused by the failure frequencies of the joints and terminations and the fact that two individual cables are used in the studied Randstad380 configuration. The difference in the unavailabilities is larger as the repair time of UGC circuits is about hundred times the repair time of OHLs. Double-circuit failures were studied using a dependent failure factor. For OHLs and shorter UGC connections, double-circuit failures are mainly dependent failures. For longer UGC connections (than about 110 km), independent double failures will happen more often than dependent double failures, depending on the connection length.

The reliability of underground cables can be improved in several ways. First, the failure frequencies and repair times can be reduced. This could be done by advanced testing and monitoring techniques and optimization of the repair process. As the repair time is much larger than the repair time of overhead lines, much improvement can be obtained. Activities like getting permission and arranging technical personnel and the availability of spare or new parts can be arranged beforehand, such that the repair time becomes significantly shorter. A second option to improve the reliability is partial cabling, where cables are only installed at those locations in a connection where they are the most desired. Studies show that short line sections (shorter than 3.6 km in the studied situation) should be avoided then. The third option is to use another cable configuration. The use of an additional spare cable theoretically gives the best improvement, but this configuration will cause practical challenges with cross-bonding, testing, and additional costs. By installing additional circuit breakers or disconnectors, it becomes possible to disconnect half a circuit after a cable failure. It must however be studied whether a 75% capacity state is acceptable in the operation of the power system, as the division of the load flow over the remaining cables depends on the impedances of the connected overhead lines in the case of an embedded cable section.

A study on the network connecting Maasvlakte380 substation showed that the reliability of the network strongly depends on the location of the UGC connections. For the probability of an islanded substation, it is better to avoid the connection of a (critical) substation by UGCs only. The probability of an overloaded circuit depends on the load flow in the network, and an UGC in a heavily loaded connection has a much higher impact than an UGC in a less heavily loaded connection. Furthermore, there is a large difference in the reliability between UGCs within one connection route and multiple UGCs within different connection routes connecting a large-scale generation center. It is found that one long UGC within one connection route can still be better than two short UGC sections within different connection routes.

A study on the risk of further cabling in the Dutch EHV transmission network showed that various factors affect the final reliability of the network. The actual loading of a connection influences this reliability significantly. In some cases, the application of series impedance compensation can cancel out the reliability impact of underground cables completely. Other influencing factors are the failure frequencies, the repair time, the total cable length, and the number of UGC sections in a partially cabled connection.

All the studies in this chapter showed that, generally, underground cables negatively influence the reliability of EHV transmission networks. The studies also showed that there is not one single reliability indicator that completely describes the reliability of a transmission network. Not only the probability of load curtailment, but also the probability of generation redispatch can be important reliability indicators for a TSO. The location of an UGC within the grid topology influences the network reliability considerably. Highly loaded connections contribute relatively

Table 7.9 Effectiveness of various design measures

Design measure	Impact
Avoid underground cables in heavily loaded connections	● ● ●
Avoid cables in multiple connection routes connecting one large generation center	● ● ●
Prevent islanding by connecting substations not only by cables	● ● ●
Apply series impedance compensation, if applicable	● ● ●
Use a configuration with disconnectors, if acceptable in system operation	● ● ●
Reduce the failure frequencies of underground cable components	● ●
Reduce the repair time of underground cable components	● ●
Use only one cable per circuit phase, compared to Randstad380	● ●
Reduce the underground cable length within a (partially cabled) connection	● ●
Install less cable sections in a partially cabled connection	●

much to the total (un)reliability, such that underground cables should be avoided in crucial, heavily loaded connections.

The possible measures and remedial actions to reduce this impact can be listed and quantified as shown in Table 7.9. The most effective measures are related to heavily loaded connections, underground cables within the same or different connection route(s), islanding of substations, and series impedance compensation. Other measures are reducing the repair time or failure frequency, using only one cable per circuit phase, and reducing the total cable length by 50%. New reliability-based design criteria can be developed for UGCs in large transmission networks from these results. As many factors are of influence on the final reliability of transmission networks with underground cables, future UGC projects should be studied individually. In this way, the possibilities to reduce the impact can be studied per individual project.

Problems

7.1 Learning Objectives
Please check the learning objectives as given at the beginning of this chapter.

7.2 Main Concepts
Please check whether you understand the following concepts:

- Cable circuit;
- Cable part;
- Joint;
- Termination;
- Repair process of underground cables;
- Partial cabling;
- Embedded cable circuit;

- Dependent failure factor;
- Cable circuit configuration;
- Series impedance compensation;
- Islanding of substations.

7.3 Underground Cable Failure Statistics

- What are the challenges regarding failure statistics of EHV underground cables?
- Mention some sources of underground cable statistics.
- How were the failure frequency and repair time of EHV underground cables determined?
- What is causing the long repair time of underground cables?

7.4 Underground Cable Circuit Reliability

- How can the failure frequency and unavailability of an UGC circuit be calculated?
- How can the (independent) failure frequency and unavailability of an UGC double circuit be calculated?
- How can the (dependent) failure frequency and unavailability of an UGC double circuit be calculated?

7.5 Underground Cable Reliability Improvement
What are the solutions to improve the reliability of an underground cable circuit?

7.6 Underground Cables in Large Transmission Networks
What can be concluded from the studies of EHV underground cables in large transmission networks?

References

1. TenneT TSO, Ministries of EZ and I&M (2015) Randstad380 project website http://www.randstad380kv.nl/
2. Wu L (2014) Impact on EHV/HV underground power cables on resonant grid behavior, PhD thesis. Technical University Eindhoven, Eindhoven, the Netherlands
3. Hoogendorp G (2016) Steady state and transient behavior of underground cables in 380 kV transmission grids, PhD thesis. Department of Electrical Engineering, Electrical Sustainable Energy, Delft University of Technology, Delft, The Netherlands
4. Meijer S, Jong JPW de, Smit JJ, Tuinema BW, Lugschitz H, Svejda G, Klein M, Fischer W, Henningsen CG, Gualano A (2012) Availability and risk assessment of 380kV cable systems in transmission grids. 44th Cigré Session, Paris, France
5. Kandalepa N, Tuinema BW, Kuik R (2014) Underground cables in the Dutch 380 electricity grid - risk assessment of further 380kV cabling. Department of Electrical Engineering, Delft University of Technology, Delft, the Netherlands, Internship Report
6. Kandalepa N, Tuinema BW, Rueda JL, Meijden MAMM van der (2015) Reliability modeling of transmission networks: an explanatory study on further EHV underground cabling in the Netherlands, MSc Thesis. Department of Electrical Engineering, Delft University of Technology, Delft, the Netherlands

7. Kandalepa N, Tuinema BW, Rueda JL, Meijden MAMM van der (2016) Reliability modeling of transmission networks: an explanatory study on further EHV underground cabling in the Netherlands. In: IEEE international energy conference, EnergyCon2016, Leuven, Belgium

8. Tuinema BW, Gibescu M, van der Meijden MAMM, van der Sluis L (2012) Reliability evaluation of underground cable systems used in transmission networks. In: 12th international conference on probabilistic methods applied to power systems (2012), PMAPS'12. Istanbul, Turkey

9. Tuinema BW, Rueda JL, van der Sluis L, van der Meijden MAMM (2015) Reliability of transmission links consisting of overhead lines and underground cables. IEEE Trans Power Deliv 31(3):1251–1260

10. Tuinema BW (2017) Reliability of transmission networks: impact of EHV underground cables and interaction of onshore-offshore networks, PhD thesis. Department of Electrical Engineering, Electrical Sustainable Energy, Delft University of Technology, Delft, The Netherlands

11. TenneT TSO (2011) NESTOR (Nederlandse Storingsregistratie) database 2006–2010. TenneT TSO B.V, Arnhem, Netherlands

12. Meijer S, Smit J, Chen X, Fischer W & Colla L (2011) Return of experience of 380 kV XLPE landcable failures. In: Jicable - 8th international conference on insulated power cables, Versailles, France

13. Meijer S, Smit J, Chen X & Gulski E (2011) Monitoring facilities for failure rate reduction of 380 kV power cables. In: Jicable - 8th international conference on insulated power cables, Versailles, France

14. TenneT TSO BV (2008) Visie2030. TenneT TSO B. V., Arnhem, the Netherlands

15. d-maps.com (2018) Free map of the Netherlands. http://d-maps.com/carte.php?num_car=15029&lang=en

Chapter 8
Cyber-Physical System Modeling for Assessment and Enhancement of Power Grid Cyber Security, Resilience, and Reliability

Power grids are increasingly dependent on operational technologies (OT) and information and communication technologies (ICT) networks for real-time monitoring and control of physical facilities. ICT and OT systems are coupled with the power grid and together, they form an interdependent and complex cyber-physical energy system (CPS). Utility ICT and OT systems have evolved from isolated, monolithic structures to open, networked environments vulnerable to cyber attacks. Cyber security therefore is an emerging research topic. Without consideration of cyber security, the digitalization of future energy systems may not be possible. Cyber attacks can damage generators and insulation of power equipment and lead to a loss of load. Cyber attacks can also affect power system dynamics and initiate cascading failures. Without proper control, they may cause a power system blackout. In this chapter, cyber security and resilience of power grids are discussed. The security controls and vulnerabilities of power systems are reviewed. Recent cyber intrusions and attacks confirmed the importance of cyber security and resilience of power grids to cyber attacks. A research program and an integrated CPS model are presented to assess and improve cyber security and resilience to cyber attacks and natural disasters of integrated cyber-physical energy systems. The power grid dynamic models are integrated with large, realistic ICT-OT communication models. Cosimulation of power systems and communication networks enables an in-depth investigation of how cyber attacks at the cyber system layer impact business continuity of utilities and the power grid at the physical layer. A CPS testbed combines the functionalities of control centers of the future and a cyber range. It is used to develop and demonstrate computational and artificial intelligence-based solutions for the enhancement of cyber security and resilience of power grids. The cyber attacks conducted against the power grid in Ukraine on December 23, 2015, are analyzed. The IEEE 39-bus system is used to

---

This chapter has been written by dr.ir. Alexandru I. Stefanov (assistant professor, Intelligent Electrical Power Grids (IEPG), dept. of Electrical Sustainable Energy (ESE), faculty of Electrical Engineering, Mathematics and Computer Science (EEMCS), Delft University of Technology, the Netherlands).

investigate how this type of targeted cyber attacks affects power system stability, causes cascading failures, and leads to a blackout.

Learning Objectives

After reading this chapter, the student should be able to:

- describe the concepts of cyber-physical system, cyber security and resilience to cyber attacks of power grids and place them in the context of recent developments and future challenges in power system digitalization;
- model an integrated cyber-physical power system and design the architecture of a testbed for assessment and enhancement of power grid cyber security, resilience and reliability;
- evaluate the impact of cyber attacks on power system dynamics and explain how cyber attacks initiate cascading failures and lead to a blackout.

8.1 Acronyms

AES	Advanced Encryption Standard
BCU	Bay control unit
CPS	Cyber-physical system
DES	Data Encryption Standard
DMS	Distribution management system
DSO	Distribution System Operator
EMS	Energy management system
EV	Electric vehicle
FIFO	First in, First out
GIS	Geographic information system
HILF	High impact low frequency
HMI	Human–machine interface
HV	High voltage
ICCP	Intercontrol Center Communication Protocol
ICS-CERT	Industrial Control Systems Cyber Emergency Response Team
ICT	Information and communication technologies
IED	Intelligent electronic device
IoT	Internet-of-Things
IP	Internet Protocol
LAN	Local area network
LV	Low voltage
MV	Medium voltage
OMS	Operations management system
OPC	Object linking and embedding for process control

OT	Operational technology
PDC	Phasor data concentrator
PLC	Programmable logic controller
PMU	Phasor measurement unit
RSA	Rivest–Shamir–Adleman
RTDS	Real-time digital simulator
RTU	Remote terminal unit
SAS	Substation automation system
SCADA	Supervisory control and data acquisition
SCS	Station control system
SQL	Structured Query Language
TSO	Transmission System Operator
VPN	Virtual private network
WAMS	Wide area monitoring system

8.2 The Cyber-Physical Power System

The energy system transition toward a sustainable, low carbon future is led by three main drivers: *decarbonization*, *decentralization*, and *digitalization*. *Integration of digital technologies* is of critical importance for the development of future electrical energy systems, e.g. Internet-of-Things (IoT), digital substations, computational and artificial intelligence techniques, and Big Data analytics [1]. The modern society heavily relies on smart electrical energy systems to supply clean and sustainable electricity, while the power grid is the backbone of *smart cities* [2]. Their integration is represented in Fig. 8.1. Active and flexible distribution networks provide electricity to millions of homes. These are connected to the transmission and subtransmission power systems and deliver electricity through the medium voltage (MV) and low voltage (LV) networks. The *smart distribution networks* also integrate electric vehicle (EV) charging infrastructures, electric transportation, energy storage, heat pumps for electric heating of smart buildings and homes, and renewable and distributed energy sources, such as solar PV, distribution-connected wind farms, and microgeneration. The *smart grid* is based on networks of connected grid edge devices and sensors, information and communication technology (ICT) and operational technology (OT) infrastructures, control centers, smart applications for monitoring and control, and data analysis capabilities. Other critical services of smart cities depend on the secure and reliable supply of clean electrical energy, i.e. healthcare, water quality, public safety and security, mobility, economic development, and IoT.

Power grids have been identified as a critical infrastructure for the modern society. They are increasingly dependent on OT systems for real-time monitoring and control of physical facilities. The OT system gathers data from substation bays and station control systems, e.g. voltage and current magnitudes, active and reactive powers, circuit breaker status, and transformer tap positions, and sends them to the control center. This real-time data is used for power system operation. Furthermore,

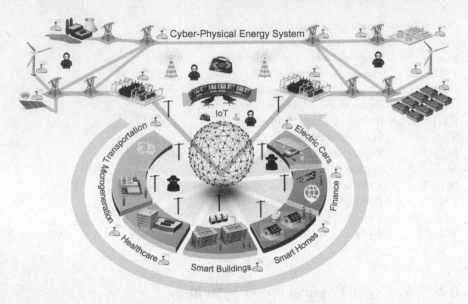

Fig. 8.1 Cyber-physical energy system for smart cities

data is processed by the *energy management system* for secure system operation, maintenance, planning, and future grid development. System operators control the physical facilities in real time by sending control commands to *remote terminal units* or *station control systems* at the substations. Utilities such as Transmission System Operators (TSOs) and Distribution System Operators (DSOs) use *ICT networks* for the non-operational side of the business, e.g. the integrated single electricity market, asset and resource management, geographic information system, finance, human resources, and payroll. The ICT and OT systems are interconnected. However, security controls such as firewalls are in place to keep the ICT-OT networks segmented and OT network separate from a direct connection to *the Internet*. It can be imagined that on top of the power infrastructure resides *layers of OT and ICT networks* that are coupled with the power grid. Together they form an interdependent and complex *cyber-physical energy system (CPS)* [3].

Reliability and *physical security* of power grids have been widely recognized for many decades as important issues and resources have been allocated to minimize component failure rates and restrict access to power system facilities. However, two new issues have recently emerged in the context of the rapid development of digital technologies: digitalization of energy systems and climate change. *Cyber security* and *resilience* of power grids to major disturbances such as cyber attacks and natural disasters are critical issues that require in-depth research to achieve adequate levels for future energy systems. *Cyber security* is the ensemble of best practices, processes, and technologies designed to protect ICT and OT infrastructures, devices, applications, and data from unauthorized access and damage while preserving the confidentiality, integrity, and availability of information and services. *Resilience* of power grids is the ability of the system to anticipate major disturbances, like cyber attacks and extreme weather events, absorb the shocks, easily adapt to new conditions and rapidly recover from disruptive events.

Cyber security, resilience, and reliability of CPS are of paramount importance for the national security. The utility ICT and OT networks have evolved from isolated, monolithic structures to open, networked systems. The *physical power and cyber layers* have become interdependent. Disruptions in the OT system because of cyber intrusions, increased latencies and communication failures impact directly the power grid monitoring and control capabilities. Major power system disturbances, such as natural disasters, may lead to power system instability, which requires proper control. OT disruptions can impede power system control and implicitly the damping of system oscillations that may lead to cascading failures. Furthermore, cascading events can also be initiated by *cyber attacks* and multiple *cyber-power component failures* leading to a blackout. The energy storage devices, like batteries and backup power supplies such as diesel generators, have limited capabilities. In case of an extended blackout, the backup power supplies will eventually run out. Without power, hospitals and other critical services are severely affected. A disruption of the cyber-physical energy system may lead to financial loss, damages, chaos, or even a loss of lives [4].

8.3 Cyber Security and Resilience of Power Grids

Cyber security is an emerging topic as ICT and OT connectivity of power grids has rapidly increased. Without consideration of cyber security, the digitalization of future energy systems is not possible. *Communication protocols* based on TCP/IP, e.g. IEC 61850, IEC 60870-5-104, and DNP3.0, and Ethernet are widely used for substation automation and communications at the bay, substation, and control center levels. *Utility private communication networks* typically use polling radio, microwave, satellite, and fiber optic infrastructures. *OT components* are networked across wide areas, i.e. bay control units, protection relays, remote terminal units, station control systems (SCS), communication servers, real-time databases, supervisory control and data acquisition (SCADA) systems, operation management systems, energy management systems, and advanced distribution management systems. Furthermore, IoT and machine-to-machine communications over *public carrier networks* are frequently used for monitoring and control of field devices such as pole-mounted autoreclosers in the MV network.

8.3.1 Security Controls for Power Grids

It is well recognized that such technologies are susceptible to IP-based *cyber attacks*. The security controls presented in Fig. 8.2 are commonly used to protect OT systems against cyber intrusions, i.e. firewalls, intrusion detection systems, virtual private

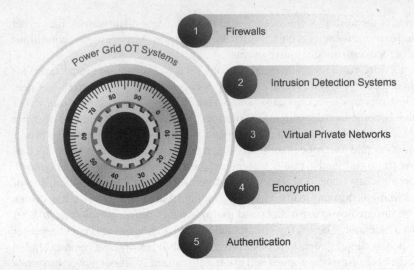

Fig. 8.2 OT security controls for power grids

networks, encryption, and authentication. *Firewalls* hold the first line of defense against cyber intrusions. Firewalls, together with routers and switches, are widely used for network segmentation and protection against unauthorized traffic from outside the utility. The network traffic is inspected using rule sets allowing legitimate data packets to pass and blocking the rest. *Intrusion detection systems* monitor the OT network for malicious activities or policy violations. Malicious activities are flagged by using either signature-based or anomaly-based detection. Event correlation methods are used for detailed analysis of patterns, data logs, and network traffic. Patterns of unexpected behavior are correlated in the signature-based detection, while deviations from network traffic considered normal are monitored and reported in the anomaly-based detection. *Alarm filtering methods* are employed to differentiate between false alarms and cyber intrusions.

Virtual private networks (VPNs) are encrypted connections between devices and/or networks over the Internet. Traffic on the virtual private network is encrypted and transmitted through a tunnel to prevent eavesdropping. *Remote access VPNs* are commonly used by OT vendors to securely connect to the utility's private communication network from remote locations over the Internet. VPN conducts security checks on endpoints before the connection is enabled. *Site-to-site VPNs* establish secure connections between utility networks separated by large distances. *Cryptographic schemes* are used to protect information and preserve data confidentiality, integrity, non-repudiation, and authentication. Cryptographic algorithms are based on mathematical concepts and computer science that allow only the sender and intended recipient to understand the message contents. Generally, private or public keys are used by the sender to encrypt data and intended recipient to decrypt. The *private-key/symmetric-key* encryption creates block ciphers with a secret key that both the sender and intended recipient use to encrypt and decrypt the message. Examples of symmetric-key encryption algorithms are the Advanced Encryption Standard (AES) and Data Encryption Standard (DES). The *public-key/asymmetric-key* encryption uses a pair of public and private keys to encrypt and decrypt data. The public key is shared with everyone, while the private key is kept secret. The sender encrypts the message with the recipient's public key. However, only the intended recipient can decrypt it with the corresponding private key. Rivest–Shamir–Adleman (RSA) algorithm is commonly used for asymmetric-key encryption. *Authentication* is widely used to verify the identity of a person or device that tries to access network resources. A simple authentication mechanism is to use usernames, passwords, personal identification numbers, and smart cards. More advanced authentication mechanisms are to use two-factor authentication, digital certificates, fingerprints, retinal scans, and voice pattern samples [5].

8.3.2 Power Grid Vulnerabilities

Usually, the *security controls* for OT systems have the same vulnerabilities as security controls used to protect the ICT systems in banking, financial, or commercial institutions. The main issue is that security controls of such ICT systems may be patched, and their vulnerabilities may be mitigated more often than the security controls used for OT systems of industrial control processes. The reasoning is that it is more difficult to update the OT systems of critical infrastructures as this process can affect the normal operation of physical facilities. A disruption of service such as electricity supply to customers may result in regulatory penalties and financial loss. Furthermore, extensive commissioning is needed after each update process that prolongs the voluntary outage for maintenance.

Although the OT network does not have a direct connection to the Internet, it does not mean that it is isolated from the Internet. OTs are interconnected with the ICT systems of utilities. Firewalls and electronic security perimeters do not guarantee cyber security. *Misconfiguration* is a common vulnerability. Furthermore, they cannot defend against *insider attacks* and *attacks originating from trusted sources*. Utility ICT systems use servers and workstations with Windows operating systems, common software applications, e.g. MS Office 365, SAP, and Citrix, and Internet connections for business-as-usual operations. Furthermore, OT vendors have *remote access* capabilities from the Internet, through the utility ICT systems, and into the OT system for software updates and technical support. The vulnerabilities of utility ICT systems may be exploited to hack into the OTs of power grids.

OT devices can be infected with *malware* such as viruses, worms, and Trojan horses that can disrupt their normal operation, damage the OT infrastructure, and corrupt real-time databases. *Denial-of-service (DoS)* attacks affect data processing through resource exhaustion such as buffer overflows and bandwidth consumption. A *cyber intrusion* into the OT system would have devastating consequences that can result in unauthorized control of the power grid, disconnection of power plants and substations from the system, change of control settings, and altering control setpoints sent to transformers and generators. Cyber attacks can lead to excessive torques that damage generators in power plants, overvoltages that damage insulation of power devices, and loss of load. Cyber attacks can also affect the voltage and rotor angle stability of the power system leading to instability. Simultaneous, multiple generator and line contingencies initiate cascading failures in the power grid and cause a power system collapse. *Blackouts* in large interconnected power systems are catastrophic failures with a severe socioeconomic impact. A comprehensive classification and analysis of previous major grid blackouts worldwide was reported [6]. For example, several blackouts occurred in 2003 in Europe and USA affecting 112 million people.

Major cyber intrusions and attacks have confirmed the importance of cyber security and resilience of power grids to cyber attacks [5]. The impact of targeted cyber attacks against the control system of an electrical generator was demonstrated by the US Department of Energy's Idaho National Laboratory in March 2007. The 'Aurora' project shows how cyber attacks can physically damage power plants [7]. Such coordinated cyber attacks on multiple power plants across the European Union can cause a complete collapse of the interconnected power grids. In June 2010, *Stuxnet* was reported as a sophisticated malware that exploits certain vulnerabilities to infect industrial control systems, spread at high infectious rates, and impact physical facilities. The main targets are the centrifuges used in nuclear plants for uranium enrichment that has applications in nuclear reactors and weapons. The worm does not affect computers that are not involved in the uranium enrichment process. It searches if the infected computer is connected to specific *programmable logic controllers (PLCs)* manufactured by Siemens. Such PLCs are used to monitor and control the physical process of uranium enrichment in centrifuges. Stuxnet reprograms parts of the PLC code for the centrifuge to spin faster and longer and damage the enrichment process. However, the infected PLC reports to the control system that the process is progressing well. This makes detection and diagnose difficult [8]. Stuxnet was created to target specific nuclear facilities that were not connected to the Internet. However, the malware was modified and infected other industrial control systems worldwide.

Malware such as Stuxnet can be adapted to attack various power plants or control centers directly affecting the security of power grid operation. In October 2012, the operation of a power plant was affected by a *virus infection*. A technician was conducting software updates using an *infected USB drive*. As a result, ten computers in the turbine control system were infected. The power plant was not operational for three weeks leading to a considerable financial loss [9]. In February 2013, cyber attacks targeted gas compressor stations in the USA. Attackers used brute force to *bypass authentication mechanisms* and access the process control network. Fortunately, none of the cyber attacks were successful [10]. In 2014, there was an increase in the number of cyber incidents against the energy sector. The Industrial Control Systems Cyber Emergency Response Team (ICS-CERT) responded to 245 cyber attacks targeting industrial assets. Their objective was to access the business and SCADA systems. *Zero-day and web application vulnerabilities* were exploited. *Structured Query Language (SQL) injection* and *targeted spear-phishing attacks* were conducted to hack into the industrial control systems. However, the exact intrusion methods are not known because of a lack of monitoring and detection capabilities within the compromised networks.

On December 23, 2015, cyber attacks were conducted against the power grid in Ukraine. Hackers intruded into ICT and OT systems of three electricity distribution companies. Seven 110 kV and twenty-three 35 kV power substations were disconnected from the power grid for hours. The cyber attacks in Ukraine are the first publicly acknowledged cyber incidents to result in power outages that affected 225,000 customers. The hackers shut down power by using *phishing emails*, BlackEnergy 3, *VPN and credential theft*, *network and host discovery*, *malicious firmware*, and *OT hijack*. Furthermore, they modified schedules for uninterruptible power supplies, opened circuit breakers, and used KillDisk for *wiping of workstations, servers, and remote terminal units* [11]. On December 17, 2016, another cyber attack was conducted on the power grid in Ukraine. It affected the SCADA system at the transmission level targeting a single 330 kV substation. The cyber attack resulted in a power outage in the distribution network where the total unsupplied load was 200 MW. This is the first publicly acknowledged malware that targeted power systems and resulted in a power outage [12].

Such laborious cyber attacks are a real threat to power grids. They are mainly conducted by cyber attackers with considerable financial resources and time for planning. Furthermore, the hacking tools used to conduct the cyber attacks are usually available to purchase at a later date on the black market. Therefore, they may also be used by hackers with fewer resources and other objectives such as *cyber terrorism and ransom*.

8.3.3 Research on Cyber Security and Resilience of Cyber-Physical Power Systems

Research has been conducted on cyber security for power grids. However, researchers with electrical engineering background usually focus on power system models, state estimation, and false data injection without the consideration of ICT and OT models [13]. Similarly, researchers with computer science background tend to focus more on information security and security controls such as cryptography and do not consider the underlying physical infrastructure [14]. Research is needed to develop scientific foundations and innovative technologies to defend against cyber attacks and mitigate cyber security threats to cyber-physical energy systems. *Integrated cyber-physical system models and CPS testbeds* are needed to study the complex system-of-system dependencies and interactions between OTs and power grids. CPS models and testbeds provide real-time, simulated data from both cyber and power system layers. These are the basis for the development and validation of novel methods and tools for cyber security assessment and enhancement of CPS, innovative defense strategies, and mitigation techniques.

A research program at Delft University of Technology (TU Delft) is summarized in Fig. 8.3. The goal is to assess and improve the cyber security and resilience to cyber attacks and natural disasters of integrated cyber-physical energy systems. This research sets the mathematical and computational foundations and develops technologies that advance the field of resiliency and cyber security to protect interconnected power grids against major disturbances in a highly connected environment.

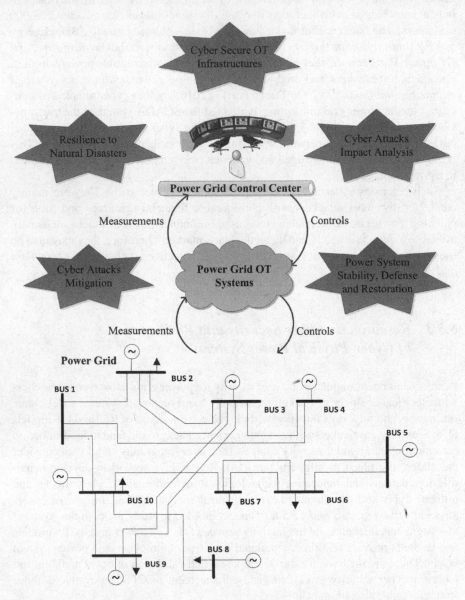

Fig. 8.3 Cyber security and resilience of cyber-physical energy systems (based on [4])

It considers future developments in transmission and distribution systems, digitalization, and cyber security threats. The objectives are to develop CPS models, computational methods, and artificial intelligence-based tools to enhance the resilience of power grids and defend against and mitigate cyber attacks as mandated by the national security strategies. The program is based on five pillars:

1. Security controls and cyber secure OT infrastructures;
2. Cyber attacks modeling, simulation, and impact analysis;
3. Cyber attacks mitigation;
4. Power system stability, defense, and restoration;
5. Resilience of power grids to natural disasters.

The ground-breaking contributions of this research and learning objectives are to:

- Model the cyber-physical energy system and integrate the power grid dynamics with large, realistic ICT-OT communication models for investigation and enhancement of cyber security and resilience to cyber attacks. The complex system-of-system dependencies are modeled and analyzed. Cosimulation of continuous-time and discrete-event systems allows in-depth investigations of how cyber attacks at cyber system layer impact business continuity of utilities and integrated energy systems at the physical layer that result in wide area outages or a blackout. The model of the integrated cyber-physical energy system provides real-time simulated data from both layers, and it is used for the research below and validation of proposed methods and tools.
- Develop methods and prototype tools to cosimulate the power system dynamics and ICT-OT communication infrastructures. The integrated simulation of cyber-physical energy systems is used for the assessment and enhancement of cyber security and resilience to cyber attacks.
- Develop testbeds that integrate real-time digital simulation of power grids with industrial-grade ICT-OT hardware, software, and communication protocols used by TSOs and DSOs for power system monitoring and control. The cyber security testbed is used to emulate security controls and cyber attacks, assess OT vulnerabilities and impact on power system operation, and test prototype technologies and rigorous defense and mitigation mechanisms needed to protect power grids. Applications such as intrusion detection, prevention and response systems, real-time, automated threat management systems, and incident response methods are developed showing their full potential of defending power grids against cyber security threats. This can be leveraged when working with hybrid cyber-power models. The findings are applied in a demonstration environment to validate the concepts and determine their real-world applicability.
- Model and simulate/emulate security controls on the testbed that protect cyber-physical energy systems and develop methods and tools for vulnerability assessment and penetration testing. Design cyber secure ICT-OT infrastructures for CPS of the future.
- Model, simulate, and emulate various cyber attacks in the CPS cosimulation environment and on the testbed, such as the cyber attacks conducted against the power

grid in Ukraine on December 23, 2015, that caused several power outages affecting 225,000 customers. Develop methods and prototype tools to:

- Assess the impact of such cyber attacks on CPS operation.
- Estimate stability margins of the power grid.
- Analyze how cyber attacks initiate cascading events that lead to a blackout.
- Define and quantify the resilience of CPS to cyber attacks.

• Develop mitigation methods and technologies to address such laborious cyber attacks based on artificial and computational intelligence, i.e. machine learning, deep learning, security games, multi-agent systems, Big Data, IoT, and edge computing, including:

- Intelligent intrusion detection, prevention and response systems;
- Real-time, automated threat management systems;
- Next-generation firewalls for CPS;
- Incident response methods and strategies;
- Tools for ICT and OT recovery planning;
- Business continuity strategies.

• Develop methods and tools for power grid defense based on:

- Intelligent decision making;
- Shock absorption and self-reconfiguration of the smart grid;
- Impact minimization and wide area emergency management;
- Stopping of cascading failures and prevention of power outages.

• Develop decision support methodologies and tools for the restoration of integrated cyber-physical energy systems. Cyber attacks can cause a blackout and damage the ICT-OT infrastructures. Computational methods and a prototype tool are needed for the efficient restoration of integrated cyber and power systems.
• Develop methods for assessment and improvement of CPS resilience to extreme weather events and natural disasters such as earthquakes, flooding, wildfire, hurricanes, and powerful storms.

8.3.4 Resilience of Cyber-Physical Power Systems to Natural Disasters

Two new types of major disturbances have emerged that can severely affect the operation and infrastructure of cyber-physical energy systems, i.e. *cyber attacks* and *natural disasters*. Extreme weather events used to be considered high-impact low-frequency (HILF) events. However, they have occurred more frequently in the past decade because of climate change with devastating impacts on critical infrastructures, for example, Hurricane Maria struck Puerto Rico on September 20, 2017. The power grid infrastructure was destroyed and about three million people were left without electricity for months [15]. Superstorms, heatwaves, and flooding can result in the sudden disconnection of generation and transmission lines and increase the risk of cascading failures and blackouts. Therefore, the power industry is transitioning from developing preventive measures to enhancing the power system operational and infrastructure resilience to major disturbances.

Resilience of critical infrastructures to extreme weather events can be assessed and improved by integrating data from various sources and using *computational and artificial intelligence* for real-time monitoring, decision making, and control. Integrated CPS models are prerequisites for the design of intelligent and resilient power grids of the future. CPS models allow the development of computational methods for the assessment of cyber-physical energy system resiliency using *Big Data and data mining techniques*, e.g. association, classification, clustering, prediction, sequential patterns, and decision trees. Such methods can be used to assess CPS vulnerabilities, likelihood and impact of extreme weather events on energy supply and demand, and quantify in real time the CPS resilience.

Furthermore, it is important to integrate the cyber and power system models to improve the operational resiliency of power grids based on *intelligent decision making* and *self-healing*. The CPS real-time data can be used to anticipate major disturbances by predicting the affected CPS areas and possible cascading failures and preparing mitigation strategies such as controlled islanding and generation redispatch scenarios that improve system stability. Special protection schemes and wide area monitoring, protection, and control can be used to stop cascading failures using high-speed OT networks. *Decision support tools* are used to reconfigure transmission and distribution systems in order to absorb the shock and minimize impact. Integrated CPS models are also important for the development of smart technologies that allow power grids to adapt to the new system condition and rapidly recover from an outage. High-speed, intelligent OT infrastructures are used to develop adaptable transmission automation technologies, distribution automation systems that detect and isolate faults, and online decision support tools for CPS condition assessment and cyber-power system restoration.

8.4 Cyber-Physical System Modeling

The power grid and communication network models are usually decoupled and studied separately. Static and dynamic grid models are used in power system analysis, while the availability and operation of communication networks between various power system controllers, substations, and control centers are assumed to be ideal. With the rapid development of ICT and OT and extensive power grid digitalization, this simplifying assumption is not valid anymore. Similarly, computer networking has been studied independently without the consideration of its application to industrial control systems and potential impact on physical processes. However, critical infrastructures such as electrical power grids are increasingly dependent on ICT and OT infrastructures. Communication failures, packet drops, and increased latencies directly impact power system operation and security of electricity supply. Cyber intrusions in the OT network and malicious cyber attacks can cripple energy systems, causing extensive power outages that lead to financial losses, equipment damage, and chaos in the modern society. Furthermore, the research to secure power grids against cyber attacks cannot be conducted effectively without the consideration of the cyber system. By focusing only on the physical layer, i.e. power system infrastructure, one can only make assumptions about the cyber system, which may fail to capture the reality. Moreover, such research would be limited to a restricted number of attack scenarios, sometimes unrealistic, such as false data injection to disrupt operations in the power grid control center.

Integration of power grid dynamic models with large, realistic ICT-OT communication models is essential for the investigation and enhancement of cyber security and resilience to cyber attacks. A comprehensive and integrated cyber-physical energy system model allows the modeling and analysis of complex system-of-system dependencies. Cosimulation of continuous time used for power system analysis and discrete-event systems for communication networks enables in-depth investigations of how cyber attacks at the cyber system layer impact business continuity of utilities and integrated energy systems at the physical layer that may result in wide area outages or a blackout. The model of the integrated cyber-physical energy system provides real-time, simulated data from both layers, and it is used for cyber security research and validation of proposed methods and tools such as intrusion detection and prevention systems and incident response strategies. Therefore, it is critical to develop methods and prototype tools to cosimulate the power system and ICT-OT communication infrastructures.

In this chapter, an integrated CPS model is proposed for assessment and enhancement of power grid cyber security, resilience, and reliability. Figure 8.4 represents the CPS layered architecture for a typical power grid. This includes four layers, i.e. power system, substation OTs, distribution control centers, and Transmission System Operator. The power system dynamic model is used to compute time-domain simulations, e.g. electromechanical transients, for analyzing power system stability (like voltage, frequency, and rotor angle stability). The time-domain simulations are synchronized with the operating system clock and computed in real time. For clarity, the computation of electromechanical transients is not accelerated, e.g. 30 s real time is needed to compute the power system dynamics over a simulated time frame of 30 s. The continuous-time electrical parameters such as voltage and current are sampled.

The power system layer exchanges simulated data in real time with the cyber system layer during the simulation of power system operating condition. Power grid measurements are computed and reported to the cyber system layer, e.g. voltage and current magnitudes, active and reactive powers, frequency, and status of circuit breakers. If phasor measurement units are considered, the measurement sets include the voltage and current phasors. Control setpoints are received from the cyber system layer, e.g. open circuit breaker, increase power generation or modify voltage setpoint, and implemented by controllers at the power system layer during simulation runtime. The dynamic behavior of the power grid is computed and reported in a close loop.

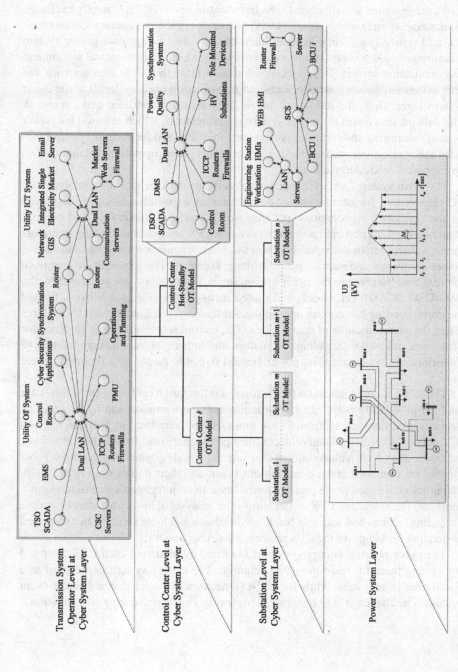

Fig. 8.4 Cyber-physical system model (based on [3, 16])

The substation level at cyber system layer includes the OT models of transmission and distribution substations. The communication processes at the bay and station levels are modeled and simulated. The bay control units (BCUs) directly exchange measurements and control setpoints with power devices, e.g. generators, transformers, and power lines. They generate data packets that encapsulate power system measurements. Bay control units are networked with the station control system and communication servers. Data packets are routed through a wide area network and sent to the distribution control centers and transmission operator levels at the cyber system layer. Here, the utility OT and ICT systems are modeled and simulated. The data packets report the power system measurements, which are used for power system monitoring and control. Furthermore, they are used for various system functions and applications such as energy management system, operational planning, grid development, and energy market.

The cyber system layer exchanges simulated data with the real-time database of a SCADA system. Measurements and circuit breaker status information are displayed on the operators' workstations and used for power grid monitoring and control. Data is communicated bidirectionally between CPS components. System operators initiate control setpoints to modify power system configuration and control the active power–frequency and reactive power–voltage balances. The setpoints are sent from the real-time database to the cyber system layer. Data packets are generated by the TSO/DSO SCADA OT models and routed through the wide area network toward the corresponding bay control units at the substation level. Latency and OT failures affect the communication of data packets. The control setpoints are sent to generator governors and automatic voltage regulators and integrated into the computation of time-domain simulations. The power system dynamic response is reported in real time.

The ICT-OT communication processes and information exchange capabilities are enabled to evaluate power grid communications performance and cyber security. Security controls such as firewalls are modeled and simulated to keep the OT and ICT networks segmented and protect utility private networks from cyber intrusions. However, they have vulnerabilities that are exploited by attackers. Various cyber attacks are simulated at the cyber system layer, and their impact on power system dynamics is assessed at the power system layer in an integrated cosimulation environment. The integrated CPS model allows the analysis of how cyber attacks initiate cascading failures and lead to a blackout. Methods and tools can now be developed to quantify and improve the CPS resilience to cyber attacks.

The cyber-physical energy system is obtained by modeling each CPS layer and integrating them for real-time data exchange. The energy system is modeled as a continuous-time system, while the ICT-OT network is modeled as a discrete-event system. The challenge is to integrate them for data exchange during cosimulation.

8.4.1 Power System Layer

The standard AC steady-state and dynamic models are used for representing power system components such as synchronous generators, power electronics interfaced generators, governors, automatic voltage regulators, power system stabilizers, under-/over-excitation limiters, on-load tap changer transformers, and dynamic loads. The power system is usually represented by a bus-branch model, where each substation is a single bus with respect to the nominal voltages. However, the bus-branch model does not allow the modification of substation breaker configuration during simulation runtime. Therefore, substations are represented for CPS modeling using the node-breaker model including circuit breakers, switches, protection relays, and current and voltage measurement transformers.

The substation model and measurement locations for the one-and-a-half breakers scheme are represented in Fig. 8.5. The scheme has two busbars, twelve circuit breakers, and eight circuits, i.e. transformer, load, and six power lines. Lines 6 and 8 are not energized as circuit breakers n_{B18}, n_{B78}, n_{B26}, and n_{B56} are open.

Sets of state and control variables are defined for data exchange with the cyber system layer. The state variables are reported to the bay control units and used for substation monitoring, e.g. measurements of voltage U [kV], active power P [MW], and reactive power Q [MVAr]. They result from the computation of time-domain simulations. Control variables are received from bay control units and used for substation

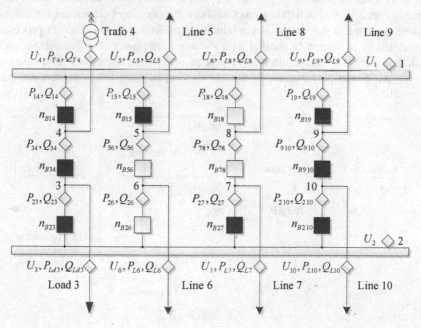

Fig. 8.5 Substation measurement model for the one-and-a-half breakers scheme

control, e.g. setpoints for circuit breaker position n_B and tap position of the power transformer. They are sent by system operators during simulation runtime. Simulated data packets communicate the measurements and control setpoints between substations, distribution control centers and Transmission System Operator. The state and control variables facilitate power system observability and control. For example, system operators can send control commands to all twelve circuit breakers during time-domain simulations and modify the substation configuration, e.g. energization of power lines 6 and 8. The measurements report the dynamic response of the power grid during power system operation.

8.4.2 Cyber System Layer

Various methods can be used to model the communication processes and OT infrastructure at the cyber system layer. One approach is to use queuing theory [17], to model the buffering, processing, and data exchange capabilities of each OT device. The OT devices are connected to create local area networks at the substation, control center and Transmission System Operator levels, and the wide area OT infrastructure.

Figure 8.6 presents the basic template for an OT model based on queuing theory. Data packets are received at a λ [packets/second] rate from the power system layer or other OT devices. The buffering capability of the OT device is modeled with a first in, first out method for organizing and manipulating the data buffer with a finite queuing capacity K. Each data packet waits in the queuing buffer a certain amount of time until one of the servers is available to process the packet. The OT processor is modeled by m servers in parallel with a computing power of ρ [packets/second]. Each data packet is delayed in the OT processor another amount of time based on the computational performance of the OT device. In case, the m servers are busy

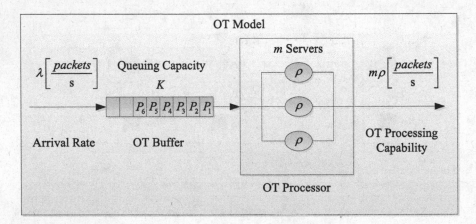

Fig. 8.6 OT model based on queuing theory (based on [3, 16])

processing data packets and the buffer is full, and incoming packets are dropped. The OT processing capability is quantified in $m\rho$ [packets/second].

The OT template is used to model each OT device at the cyber system layer represented in Fig. 8.4. The overall communication performance depends on the buffering and processing capabilities of each component and the networking configuration. Increased latency, data packet loss, and throughput affect OT communications and power system monitoring and control. Furthermore, malicious data packets can be generated by cyber intruders to disrupt the OT infrastructure via denial-of-service attacks. An avalanche of incoming data packets would rapidly overflow the OT buffer, consume its bandwidth, and affect the processing capability. As a result, power system data is lost and the OT device would fail to perform its functions. Furthermore, confidential data can be accessed and modified in the man-in-the-middle attack. Cyber attacks can capture and alter data packets carrying legitimate setpoints to power system controllers. The alteration of setpoint integrity can directly affect power system stability. Another cyber attack scenario is the compromise of OT devices such as station control systems and bay control units. The attackers send unauthorized control setpoints to operate circuit breakers, modify generator power and voltage settings, change configuration of protective relays, and damage the OT infrastructure.

The bay control unit is an OT that interacts with power devices at the substation bay level. Figure 8.7 represents the OT model for BCU. Power grid measurements are made available at the bay level during time-domain simulation at the power system layer. The state variables are linked with the digital input ports of BCU. For example, the bus voltage measurement corresponds to digital input 1. Voltage is measured with a predefined sampling rate. An event-based entity generator creates a discrete event when a new measurement is provided. Each voltage measurement is encoded into a data packet together with various attributes, e.g. IP destination and source addresses. The data packets are combined on the same path using the Round Robin algorithm. Packets are stored in the BCU buffer and processed. They are outputted on the local operating network and communicated with the station control system. Bidirectional communications are enabled. The station control system sends data packets with control setpoints to the bay control unit. Packets are stored in the BCU memory and processed. The payloads, i.e. setpoints, are extracted, sent to the corresponding digital output port, and the discrete events are discarded. For example, voltage setpoints are extracted and communicated to digital output port 1 to be actuated. The voltage control setpoints are sent to the automatic voltage regulator simulated at the power system layer. The automatic voltage regulator implements the new setpoints in the control system of a power generator or transformer.

Figure 8.8 represents the substation OT model. Each power system bay has a bay control unit with sensing, actuating and protection capabilities. The BCUs are networked with the station control system over the local operating network using fiber optics. The SCS provides all monitoring and control capabilities for modern substations. It communicates with the substation real-time database, human–machine interfaces (HMIs) and communications server. A router/firewall is used to enable remote communications with the distribution control center/Transmission System Operator using fiber optics. Alternatively, for older substations, polling radio, microwave, and

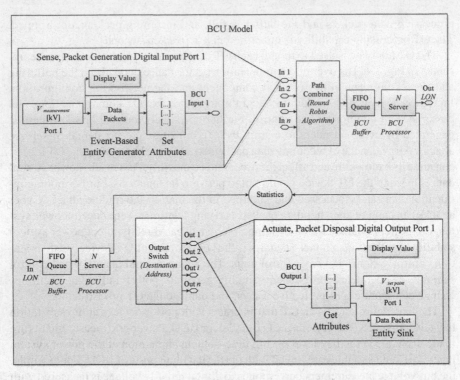

Fig. 8.7 OT model for BCU (based on [3, 16])

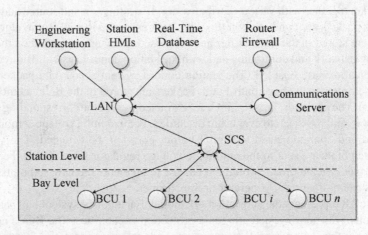

Fig. 8.8 Substation OT model (based on [3, 16])

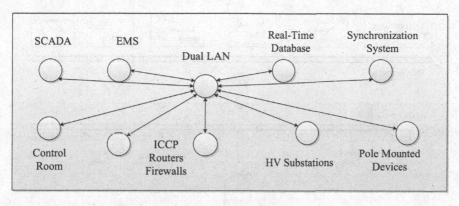

Fig. 8.9 Control center OT model (based on [3, 16])

satellite are used to communicate real-time data such as measurements and controls with the upper supervisory levels.

Figure 8.9 represents the typical OT model of a control center. The most critical OT component is the SCADA system. This communicates remotely with all station control systems, remote terminal units, and pole-mounted devices in the area of control. Power system measurements are gathered at various update rates, e.g. one second, by polling remote terminal units and SCS. They are stored in a real-time database and displayed on the single line diagram of the power grid in the control room. System operators use the single line diagram and real-time measurements for power system monitoring and control. Furthermore, power system data is processed by various applications in the energy management system, e.g. state estimation, power flow, and contingency analysis.

Typically, the high voltage (HV) and extra high voltage (EHV) substations are controlled by the Transmission System Operator (TSO), while the substations at subtransmission and medium voltage levels are supervised by Distribution System Operators (DSOs). The TSO license allows active power–frequency control system wide and reactive power–voltage control locally at the transmission system level. The DSO license allows only reactive power–voltage control locally at the distribution system level.

Figure 8.10 represents the ICT-OT models for Transmission and Distribution System Operators. The TSO and DSO operational technologies, i.e. TSO and DSO SCADA LANs, have similar components and functionalities for transmission and distribution system monitoring, control, planning, and development. The distribution and transmission control centers exchange operational information using the Inter-control Center Communication Protocol (ICCP). Usually, the TSO and DSOs are separate utilities. Therefore, firewalls are used on both sides to protect the utility of private communication networks. In each utility, the OT network is segmented from the corporate ICT network for cyber security considerations. The traffic between the OT-ICT networks is controlled. On the TSO corporate side, the ICT system is used for network mapping in the geographic information system (GIS), asset and resource

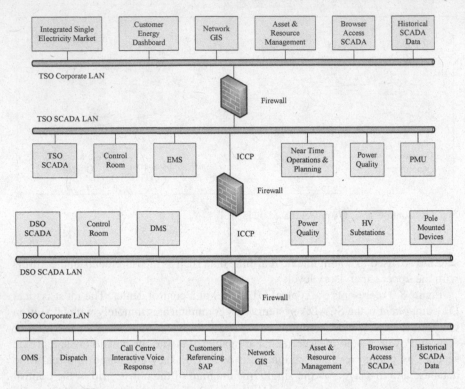

Fig. 8.10 ICT-OT models for TSO and DSO

management, and power grid development planning. The SCADA data from the OT system is used by the integrated single electricity market and customer energy dashboards. Similarly, on the DSO corporate side the ICT system is used to report and detect faults in the operations management system (OMS) and to dispatch field crews to restore power to electricity customers.

8.5 Cyber-Physical System Testbed

Development and implementation of computational and artificial intelligence-based solutions for cyber security and resilience of power grids can be conducted and demonstrated in realistic environments such as a cyber-physical energy system testbed. A CPS testbed is a setup based on industrial-grade ICT-OT hardware and software that emulates the electrical and communication processes in real time. It is an efficient method to conduct a broad range of cyber attacks and other major disturbances, analyze their impact on the cyber-physical energy system, and develop robust prevention and mitigation methods and technologies in a realistic, but safe environment. It is obvious that testing the impact of cyber attacks on the real-world industrial control systems is not an option. This would directly affect the physical processes such as electricity supply that we are trying to protect. Therefore, modeling and simulation tools are a viable alternative. However, in-depth simulation of complex processes may be difficult and limited by the computational burden. Cyber-physical system testbeds play an important role in overcoming the simulation challenges. Research efforts have developed powerful tools for the identification of SCADA vulnerabilities and cyber security studies [18, 19].

A cyber-physical system testbed at TU Delft combines the functionalities of a control center of the future and cyber range. The control center of the future is a realistic setup of the present control center used for research on innovative energy management systems, functions, and applications for future power grids. Novel technologies are being developed such as autonomous, self-driving, and self-healing power systems. The cyber range is a testbed used for cyber security investigations. Various targeted cyber attacks and malware are conducted against the cyber-physical energy system and evaluated in a contained, safe environment. Defense mechanisms and incident response strategies are developed and tested.

Figure 8.11 represents the architecture of the cyber-physical system testbed. It consists of two control centers for transmission and distribution systems, respectively. The control centers emulate the utility ICT and OT systems as shown in Figs. 8.10 and 8.11. They exchange information via ICCP. The control centers are networked with substations and communicate data in real time using power system communication protocols such as DNP 3.0, IEC 60870-5-101/104, IEC 61850, and IEEE C37.118. Transmission and distribution substations are emulated and include station control systems, bay control units, remote terminal units, and intelligent electronic devices, which are networked by using IEC 61850 and implemented via fiber optics. The substation operational technologies are mapped with real-time digital simulators (RTDS), DIgSILENT PowerFactory and communication network emulation, and simulation software. RTDS and DIgSILENT PowerFactory simulate various power systems such as the transmission digital twin and connected distribution systems. They compute time-domain simulations and provide real-time, simulated data with a sampling rate to the substation OT systems. The data exchange is enabled by IEC 61850, IEEE C37.118, and object linking and embedding for process control (OPC). SCS, communication protocols, and SCADA systems report the measurements to control centers that are displayed on the power system single line diagram in the control rooms.

System operators monitor and control the simulated power grid. Any control commands are sent in a realistic environment through the OT system and implemented in the power system simulators. The communication network emulation and simulation software overcome OT scalability and latency issues. They enable communications over a wide area network. The testbed includes the security controls shown in Fig. 8.2, which are used to protect the ICT-OT systems and control centers. A cyber range consists of tools and systems for both attacking and defending/incident response. Cyber attackers try to disrupt the operation of control centers, communication processes, and power system stability. Network defenders use intelligent intrusion detection,

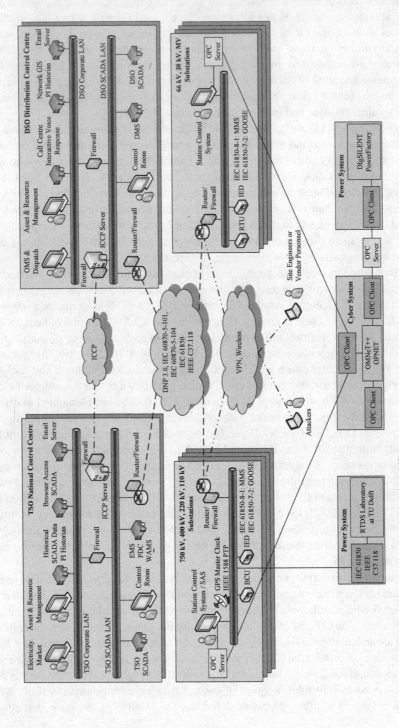

Fig. 8.11 Cyber-physical system testbed (based on [4])

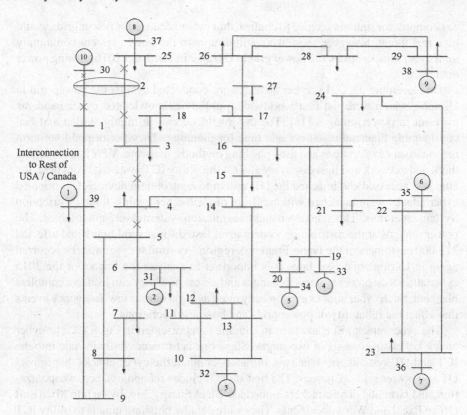

Fig. 8.12 IEEE 39-bus test system

prevention and response systems, business continuity strategies, and automated threat management systems to defend, respond to cyber attacks, and improve power grid resiliency.

For educational purposes, standard IEEE test systems are used as an alternative to transmission digital twins. The IEEE 39-bus, i.e. New-England, is a well-known and validated system [20]. It consists of 10 synchronous power plants, 39 busbars, and 46 branches as shown in Fig. 8.12. Generator 1 (bus 39) represents the interconnection to the rest of USA and Canada. Generator 2 at bus 31 is the reference machine.

8.6 Cyber Attacks on Power Grids

Research on cyber security for power grids emerged more than a decade ago. The research community highlighted cyber security as an important research issue with a potential devastating impact on power grids. Cyber security incidents against OTs for various industrial control systems such as nuclear facilities, water utilities, and gas compressor stations seemed to confirm that cyber security for power grids would be a real threat. However, there were still doubts in the power system community until cyber attacks struck the power grid in Ukraine in 2015 and 2016 causing power outages.

On December 23, 2015, cyber attacks were conducted against the power grid in Ukraine, which were the first worldwide, publicly acknowledged cyber incidents to result in power outages [11]. The cyber attackers were highly skilled and had considerable financial resources and time for planning. They performed long-term reconnaissance operations and used phishing methods, malware, VPN and credential theft, and network and host discovery against the utility ICT systems. From here, the attackers accessed and hijacked the OT system to control field devices. Coordinated cyber attacks were launched within 30 min of each other against three Distribution System Operators. The generation and transmission systems were not affected. The power outages at the distribution system level lasted for several hours and affected 225,000 customers in the Ivano-Frankivsk region. As similar cyber attacks occurred again in Ukraine one year later, it is important to analyze the impact of the 2015 cyber attacks on power systems dynamics and assess how they can lead to a complete blackout. Such cyber attacks are currently rated as high-impact low-frequency events that affect the reliability of power grids and business operations.

The cyber attack kill chain used in Ukraine is represented in Fig. 8.13. The cyber attack kill chain consists of two stages. Stage one is to successfully intrude into the ICT and OT systems. In stage two, intruders conduct the cyber attacks that impact OT and power grid operations. The first stage includes reconnaissance, weaponization, and targeting. The attackers embedded BlackEnergy 3 malware into Microsoft Office Excel and Word documents. They were sent by phishing emails to utility ICT and administrative personnel. The Excel and Word documents were opened, and popup displays instructed employees to enable document macros, which installed BlackEnergy 3 malware on the victim's system. The malware opened gateways into the infected ICT systems. The gateways allowed intruders to gather information and

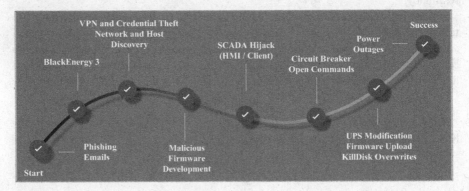

Fig. 8.13 Ukraine cyber attack kill chain

enable remote access. It appears that intruders accessed the utility ICT systems six months before December 23rd when the cyber attack was conducted. Next, intruders used the remote access to steal login credentials, escalate access privileges, discover networked systems and hosts, and establish persistent access to targets. Intruders also identified VPN connections from the utility ICT system into the OT system. The VPN theft allowed the discovery of the remainder OT systems including SCADA servers and control room workstations. Such unauthorized access level allowed intruders to thoroughly plan the cyber attack in the second stage. For example, intruders identified a network-connected uninterruptible power supply and installed malicious firmware such that the cyber attack would also result in a power outage in the utility buildings and control center. Intruders explored means to interact with the distribution management system and operator HMIs by using the existing remote administration tools. They also created malicious firmware for serial-to-Ethernet adapters used to interpret control commands from the SCADA system to substation control systems. Finally, intruders installed KillDisk malicious software across the OT system and locked system operators out of their workstations. The cyber attack was executed on December 23rd. Intruders used the HMIs of the hijacked SCADA system to take control of the distribution system and open circuit breakers. Seven 110 kV and twenty-three 35 kV substations were disconnected from the power grid for hours. At the same time, attackers installed the malicious firmware on serial-to-Ethernet gateways to ensure that Distribution System Operators cannot send remote control commands to substations and restore power. The cyber attacks caused power outages in the distribution systems [11].

The cyber attack kill chain was replicated on the cyber-physical system testbed. The main difference is that the dynamic model of IEEE 39-bus system is used instead of the Ukrainian power grid to analyze how this type of cyber attacks affect power system stability and cause cascading failures. The substations and power devices disconnected from the power system are fictitious and correspond to IEEE 39-bus test system presented in Fig. 8.12. The simulated cyber attack scenario is that intruders disconnect transformer 2–30 and power lines 1–2, 2–25, 2–3, 3–4, and 4–5 at 0, 5, 10, 15, 20, and 25 s, respectively, from the start of time-domain simulations. The affected power system components are represented in Fig. 8.12. As a result, the power system stability is affected. Figure 8.14 shows the generator rotor angles with reference to generator 2 angle, and Fig. 8.15 presents the generator speeds. At 25.872 s generator 1 is out of step and loses synchronism with the rest of the power system. One second later, at 26.872 and 26.882 s, generators 4, 5, 6, and 7 are out of step. Finally, at 26.922 and 26.932 s, generators 8 and 9 lose synchronism. Generators 2 and 3 remain in synchronism but they cannot supply the power system load. Figures 8.16 and 8.17 present the bus voltage variations. Once the generators are out of synchronism, the bus voltages are significantly affected, dropping to alarming levels below 0.8 p.u. and oscillate. Figures 8.18, 8.19, 8.20 and 8.21 show the frequency, voltage and active and reactive powers measured on the interconnector. The large oscillations indicate that the power system transient stability is affected. The protection system would act fast and disconnect the generators out of synchronism. This would also disconnect overloaded transmission lines. Cascading failures are

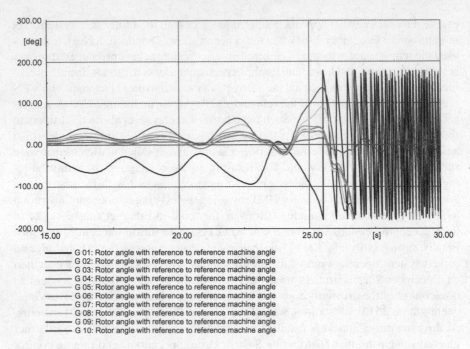

Fig. 8.14 Generator rotor angles

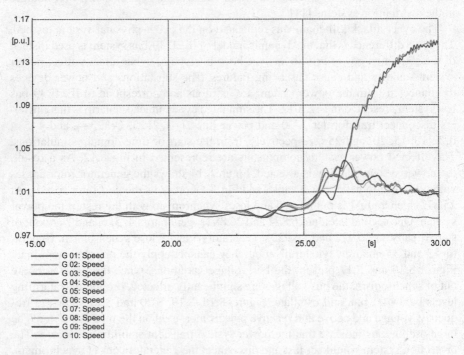

Fig. 8.15 Generator speeds

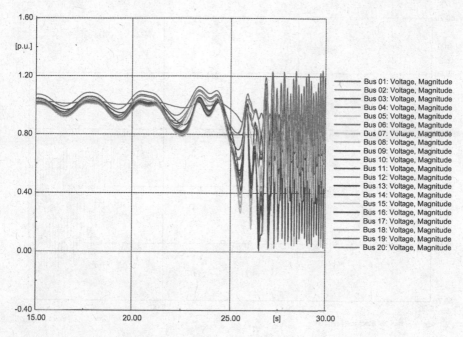

Fig. 8.16 Voltage magnitudes at buses 1–20

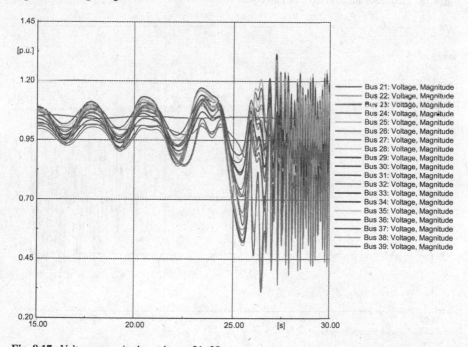

Fig. 8.17 Voltage magnitudes at buses 21–39

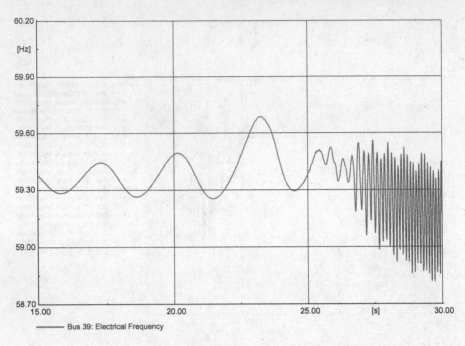

Fig. 8.18 Frequency measured on the interconnector

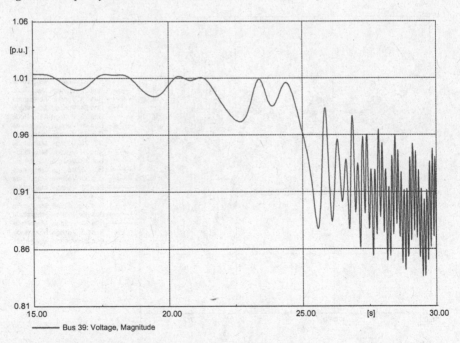

Fig. 8.19 Voltage measured on the interconnector bus

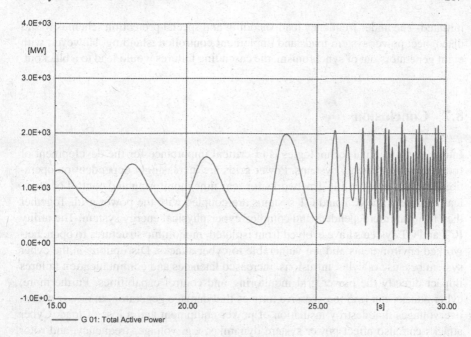

Fig. 8.20 Active power measured on the interconnector

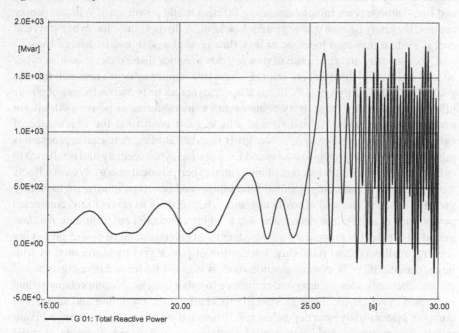

Fig. 8.21 Reactive power measured on the interconnector

initiated. The under-frequency load-shedding and special protection schemes would disconnect power system loads and implement controlled islanding. However, with eight generators out of synchronism, the cascading failures would lead to a blackout.

8.7 Conclusion

Integration of digital technologies is of critical importance for the development of future electrical energy systems. Power grids are increasingly dependent on operational technologies and ICT networks for real-time monitoring and control of physical facilities. The ICT and OT systems are coupled with the power grid. Together they form an interdependent and complex cyber-physical energy system. The utility ICT and OT systems have evolved from isolated, monolithic structures to open, networked environments and are vulnerable to cyber attacks. Disruptions in the cyber system because of cyber intrusions, increased latencies and communication failures impact directly the power grid monitoring and control capabilities. Furthermore, cyber attacks can lead to excessive torques that damage generators in power plants, overvoltages that destroy insulation of power equipment and a loss of load. Cyber attacks can also affect power system dynamics, e.g. voltage, frequency, and rotor angle, leading to instability. Multiple cyber-power component failures and generator and line contingencies initiate cascading failures in the power grid. Without proper control, they may cause a power system blackout. A disruption of the cyber-physical energy system may lead to financial loss, damages, chaos, or even a loss of lives.

Cyber security and resilience of power grids to major disturbances such as cyber attacks and natural disasters are critical issues that require in-depth research to safeguard the future energy systems. In this chapter, cyber security and resilience of power grids are discussed. The security controls and vulnerabilities of power systems are reviewed. Recent cyber intrusions and attacks have confirmed the importance of cyber security and resilience of power grids to cyber attacks. A research program is presented where the goal is to assess and improve the cyber security and resilience to cyber attacks and natural disasters of integrated cyber-physical energy systems. It sets the mathematical and computational foundations and develops innovative technologies that advance the field of resiliency and cyber security to protect interconnected power grids against major disturbances in a highly connected environment. An integrated CPS model is proposed for assessment and enhancement of power grid cyber security, resilience, and reliability. Integration of power grid dynamic models with large, realistic ICT-OT communication models is essential for the investigation and enhancement of cyber security and resilience to cyber attacks. A comprehensive and integrated cyber-physical energy system model allows the modeling and analysis of complex system-of-system dependencies. Cosimulation of power systems and communication networks enables in-depth investigations of how cyber attacks at cyber system layer impact business continuity of utilities and the power grid at the physical layer that result in wide area outages. Development and implementation of computational and artificial intelligence-based solutions for cyber security and resilience of power grids are conducted and demonstrated in realistic environments such as a cyber-physical energy system testbed. A CPS testbed at TU Delft combines the functionalities of control centers of the future and cyber range. The control center of the future is a realistic setup of the present control center used for research on innovative energy management systems, functions, and applications for future power grids. The cyber range is a testbed used for cyber security investigations where various cyber attacks are conducted against the cyber-physical energy system and evaluated in a contained, safe environment. The cyber attacks conducted against the power grid in Ukraine on December 23, 2015, are analyzed. IEEE 39-bus system is used to investigate how this type of targeted cyber attacks affects power system stability, causes cascading failures and leads to a blackout.

Problems

8.1 Learning Objectives
Please check the learning objectives as given at the beginning of this chapter.

8.2 Main Concepts
Please check whether you understand the following concepts:

- Cyber-physical energy system;
- Cyber security;
- Resilience of power grids;
- Cyber-physical system testbed.

8.3 Operational Technologies and ICTs for Power Grids

- Explain how utility OT and ICT networks are used for power system operation.
- Explain how OT and ICT disruptions and cyber attacks can impact power grids and society.
- What security controls are commonly used to protect OT systems against cyber intrusions?
- Is the power grid OT system more vulnerable to cyber intrusions that the ICT systems of other financial and commercial institutions? Please explain.

8.4 Cyber-Physical Power System Modeling

- Explain why integrated CPS models are needed for power system analysis.
- Describe the CPS layered architecture for a typical power grid.
- Explain how queuing theory is used to model the communication processes of OT devices.

8.5 Cyber Attacks on Power Grids

- What are the potential consequences of cyber intrusions into the OT systems of power grids?
- Briefly describe the cyber attacks conducted against the power grid in Ukraine on December 23, 2015.

References

1. Chen S, Wen H, Wu J, Lei W, Hou W, Liu W, Xu A, Jiang Y (2019) Internet of things based smart grids supported by intelligent edge computing. IEEE Access 7:74089–74102
2. Masera M, Bompard EF, Profumo F, Hadjsaid N (2018) Smart (electricity) grids for smart cities: assessing roles and societal impacts. IEEE Proc 106(4):613–625
3. Stefanov A, Liu C-C, Govindarasu M, Wu S-S (2015) SCADA modeling for performance and vulnerability assessment of integrated cyber-physical systems. Int Trans Electr Energy Syst 25(3):498–519
4. Stefanov A, Liu C-C (2014) Cyber-physical system security and impact analysis. In: Proceedings of the international federation automatic control (IFAC), Cape Town, pp 11238–11243
5. Xie J, Stefanov A, Liu C-C (2016) Physical and cyber security in a smart grid environment. Wiley Interdiscip Rev (WIRE) - Energy Environ 5(5):519–542
6. Besanger Y, Eremia M, Voropai N (2013) Major grid blackouts: analysis, classification, and prevention. In: Eremia M, Shahidehpour M (eds) Handbook of electrical power system dynamics, 1st edn. IEEE Press, Hoboken, pp 789–863
7. Bumgarner J (2010) Computers as weapons of war. Inf Oper J 2(2):4–8
8. Chen TM, Abu-Nimeh S (2011) Lessons from Stuxnet. IEEE Comput Soc 44(4):91–93
9. Industrial Control Systems Cyber Emergency Response Team (2012) Incident response activity. ICS-CERT monitor (December 2012). https://ics-cert.us-cert.gov/sites/default/files/Monitors/ICS-CERT_Monitor_Oct-Dec2012.pdf
10. Industrial Control Systems Cyber Emergency Response Team (2013) Incident response activity. ICS-CERT monitor (June 2013). https://ics-cert.us-cert.gov/sites/default/files/Monitors/ICS-CERT_Monitor_Apr-Jun2013.pdf
11. Lee RM, Assante MJ, Conway T (2016) Analysis of the cyber attack on the Ukrainian power grid. SANS - Industrial Control Systems, E-ISAC white paper
12. Assante MJ, Lee RM, Conway T (2017) Modular ICS malware. SANS - Industrial Control Systems, E-ISAC white paper
13. Che L, Liu X, Li Z, Wen Y (2019) False data injection attacks induced sequential outages in power systems. IEEE Trans Power Syst 34(2):1513–1523
14. Li Y, Zhang P, Huang R (2019) Lightweight quantum encryption for secure transmission of power data in smart grid. IEEE Access 7:36285–36293
15. Kwasinski A (2018) Effects of Hurricane Maria on renewable energy systems in Puerto Rico. In: Proceedings of the international conference on renewable energy research and applications, Paris, pp 1–8
16. Stefanov A, Liu C-C, Liyanage K (2015) ICT modeling for cosimulation of integrated cyber-power systems. In: Pathan A-SK (ed) Securing cyber-physical systems. CRC Press, Taylor & Francis Group, Boca Raton
17. Cassandras CG, Lafortune S (2008) Introduction to discrete event systems, 2nd edn. Springer, Berlin
18. Hahn A, Govindarasu M, Sridhar S, Kregel B, Higdon M, Adnan R, Fitzpatrick J (2019) Development of the powercyber SCADA cyber security testbed. In: Proceedings of the cyber security and information intelligence research workshop, Oak Ridge National Lab, pp 21:1–4
19. Liu C-C, Stefanov A, Hong J, Panciatici P (2012) Intruders in the grid. IEEE Power Energy Mag 10(1):58–66
20. Athay T, Podmore R, Virmani S (1979) A practical method for the direct analysis of transient stability. IEEE Trans Power Appar Syst PAS-98(2):573–584

Part IV
Conclusion

Chapter 9
Conclusion

This book gave an introduction into probabilistic reliability analysis of power systems. Various approaches to model the reliability of components, small systems and large systems were presented and illustrated by examples of real case studies. For individual components, it was discussed how their reliability behavior can be modeled by reliability functions (unrepairable components), the component life cycle (repairable components) and the two-state Markov model (repairable and unrepairable components). For small systems, it was discussed how these can be modeled by reliability networks, Markov models and fault/event tree analysis. For larger networks, the probabilistic approaches state enumeration, generation adequacy and Monte Carlo simulation were discussed.

The choice for a certain approach always depends on the characteristics of the study. Apart from the size of the studied system, the choice for a certain probabilistic approach depends on the objectives of the particular study. For example, it was mentioned that Monte Carlo simulation can be used to realistically simulate the behavior of a power system (e.g. time behavior of generation/load and switching sequences). Another example is fault/event tree analysis, which gives a clear overview of the combination of component failures (or the sequence of events) that leads to a failure of the system. In this sense, there is no approach that always works best.

The reliability of power systems can be improved by risk mitigation. Risk mitigation can be performed at several levels, but always starts with a certain awareness of the possible risks and the introduction of reliability analysis. In this way, systems are designed, implemented and operated for reliable operation. Probabilistic reliability analysis can be a useful tool to quantify and reduce the risks. Risk can then be mitigated at component and system level. At component level, risk can be mitigated by reducing the failure frequencies and repair times of the components. In practice, this can for example be accomplished by advanced testing, installation, monitoring and maintenance techniques. Repair times could be shortened by optimizing the repair process, like discussed for EHV underground cable connections. At system level, risk can be mitigated by reducing the likelihood of failure events or limiting their impact. Redundancy is an effective and widely applied risk mitigation measure, which can

© Springer Nature Switzerland AG 2020
B. W. Tuinema et al., *Probabilistic Reliability Analysis of Power Systems*,
https://doi.org/10.1007/978-3-030-43498-4_9

however be costly in cases like the connection of offshore wind farms. Limiting the impact of failure events could for example be accomplished by backup generation or demand side response.

This book also presented three topics related to probabilistic reliability analysis of power systems. The first, probabilistic load flow analysis, focused on the modeling of uncertainties in the power system, like load behavior and variable (renewable) generation. The second topic, reliability of EHV underground cables, discussed how the methods discussed in the theoretical chapters can be applied to analyze the reliability of underground cables in large transmission networks. The third topic, cyber-physical power system modeling, discussed how the reliability and resilience of the power system to cyber attacks become important in future power systems. These three topics showed how the theory of this book can be applied in practice and how the theory is related to other research fields. Cooperation with other fields of research generally helps to obtain the most reliable solution in practice.

Probabilistic reliability analysis of power systems is a developing research field. New approaches are still being introduced, while existing methods are adapted to new situations and developments. This book presented the basis of probabilistic reliability analysis, in order to provide students with the background needed to work in this field. For more detailed descriptions of the methods discussed in this book as well as new developments of probabilistic approaches, the student is recommended to consult some of the interesting textbooks and related publications on this topic. We hope that this book provided the basic knowledge that is needed to work with enthusiasm in this research field.

Appendix A
Probability and Statistics

A.1 Probability

In probabilistic analysis, one tries to predict and quantify the possible outcomes of processes that are random by nature. For example, one would like to know the probability that a transmission line fails. Even though such a process is probabilistic by nature, it is still possible to analyze it and predict the behavior to a certain extent.

In probabilistic analysis, often specific *experiments* are studied. A stochastic experiment can be described by a *random variable*, which can take on various possible *outcomes* from a set of all individual (i.e. mutually exclusive) possible outcomes called the *sample space*. Probabilities can be assigned to the individual outcomes.

A sample space can, for example, be $S_{\text{coin}} = \{heads, tails\}$ for the experiment of flipping a coin and $S_{\text{dice}} = \{1, 2, 3, 4, 5, 6\}$ for the experiment of throwing a dice. Consider two sets: $A = \{1, 2, 3, 4\}$ and $B = \{3, 4, 5, 6\}$. With the operators $\in$ (element of), $\notin$ (not an element of), $\cup$ (union), and $\cap$ (intersection), we can write:

$$1 \in A \tag{A.1}$$

$$1 \notin B \tag{A.2}$$

$$A \cup B = \{1, 2, 3, 4, 5, 6\} \tag{A.3}$$

$$A \cap B = \{3, 4\} \tag{A.4}$$

Maybe the most basic probabilistic experiment is flipping a coin. For a fair coin, the probability of the outcome *tails* is $p = 0.5$ and *heads* is $q = 1 - p = 0.5$. This is an example of a *Bernoulli experiment* [1]:

© Springer Nature Switzerland AG 2020
B. W. Tuinema et al., *Probabilistic Reliability Analysis of Power Systems*,
https://doi.org/10.1007/978-3-030-43498-4

$$P_X(x) = P_X(X = x) = \begin{cases} 1-p & x = 0 \ (\textit{'heads'}) \\ p & x = 1 \ (\textit{'tails'}) \\ 0 & \text{otherwise} \end{cases} \tag{A.5}$$

$$(\text{for } 0 \le p \le 1)$$

where:

X = random variable representing the probabilistic experiment (*flipping a coin*)

x = possible outcome

p = probability of success

$P_X(x)$ = function that assigns probabilities to the possible outcomes

A Bernoulli experiment typically leads to a *success*, with probability p, or a *fail*, with probability $1 - p = q$. A more advanced experiment can lead to more outcomes. For example, throwing a fair dice can be described by:

$$P_Y(y) = \begin{cases} 1/6 & y = 1 \\ 1/6 & y = 2 \\ 1/6 & y = 3 \\ 1/6 & y = 4 \\ 1/6 & y = 5 \\ 1/6 & y = 6 \\ 0 & \text{otherwise} \end{cases} \tag{A.6}$$

where:

Y = random variable representing the probabilistic experiment *throwing a dice*

y = possible outcome

$P_Y(y)$ = function that assigns probabilities to the possible outcomes

A graphical representation of this experiment is shown in Fig. A.1.

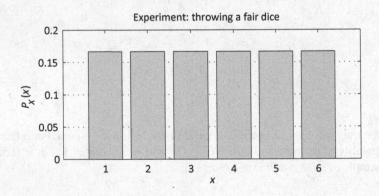

Fig. A.1 PMF of experiment: throwing a fair dice

The kind of function representing the previous two experiments, and the example illustrated in Fig. A.1, is called *probability mass function (PMF)* [1]. The probability mass function has three properties:

1. For any x, $P_X(x) \geq 0$.
2. $\sum_{x \in S_X} P_X(x) = 1$.
3. For any event $B \subset S_X$, the probability that X is in the set B is $P[B] = \sum_{x \in B} P_X(x)$, where S_X is the set of all possible outcomes.

In other words, every outcome has a positive probability (1), the total probability of all outcomes is 1 (2), and the total probability of multiple outcomes is the sum of their individual probabilities (3).

Experiments that have continuous instead of discrete outcomes are described by *probability density functions (PDFs)*. For example, the outcome of an ideal random number generator can be described by:

$$f_Z(z) = \begin{cases} 1 & 0 \leq z \leq 1 \\ 0 & \text{otherwise} \end{cases} \tag{A.7}$$

where:
Z = random variable representing the probabilistic experiment *random number generator*
z = possible outcome
$f_Z(z)$ = function that assigns probabilities to the possible outcomes

This experiment is an example of a *uniform random variable* [1]. Because the probability of a single outcome equals 0 (e.g. an ideal random number generator generating exactly 0.75), probabilities of ranges of outcomes are calculated by integration of the function instead. For example, the probability of generating a number < 0.5 can be calculated by:

$$P[X < 0.5] = \int_0^{0.5} f_x(x)\mathrm{d}x = \int_0^{0.5} 1 \, \mathrm{d}x = [x]_0^{0.5} = 0.5 \tag{A.8}$$

Also, the probability density function has three properties:

1. For any x, $f_x(x) \geq 0$.
2. $\int_{-\infty}^{\infty} f_x(x)\mathrm{d}x = 1$.
3. $P[a < X \leq b] = \int_a^b f_x(x)\mathrm{d}x$.

With the third property, a function called the *cumulative distribution function (CDF)* can be defined [1]. This CDF describes the probability that the outcome is smaller than a certain value and is related to the PDF by:

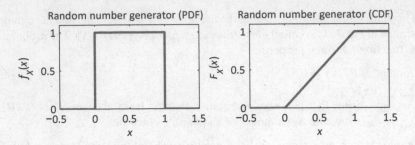

Fig. A.2 PDF and CDF for the random number generator

$$F_X(x) = P[X \le x] = \int_{-\infty}^{x} f_X(u)\mathrm{d}u \qquad (A.9)$$

Consequently: $\frac{\mathrm{d}F_X(x)}{\mathrm{d}x} = f_X(x)$.

The relation between the PDF and the CDF is also shown graphically in Fig. A.2 for the random number generator. The CDF clearly shows that the probability of $X < 0.5 = 0.5$ and the probability of $X \le 1 = 1$.

Suppose that we would like to know the probability of generating a number smaller than 0.5, given that a number smaller than 0.75 is generated. This can be calculated by the *conditional probability* [1]:

$$P[A|B] = \frac{P[A \cap B]}{P[B]} = \frac{P[AB]}{P[B]} \qquad (A.10)$$

$$P[X < 0.5 | X < 0.75] = \frac{P[X < 0.5]}{P[X < 0.75]} = \frac{0.5}{0.75} = \frac{2}{3} \qquad (A.11)$$

The *expected value* of a probabilistic function is an indication of the average outcome of the experiment. For discrete functions, it is defined by:

$$E[X] = \sum_{x \in S_X} x P_X(x) \qquad (A.12)$$

where:
S_X = the set of possible outcomes
$P_X(x)$ = the probability of outcome x
$E[X]$ = expected value of random variable X

For continuous functions, the expected value is defined by:

$$E[X] = \int_{-\infty}^{\infty} x f_X(x)\mathrm{d}x \qquad (A.13)$$

A variety of PMFs and PDFs exists to describe different stochastic experiments and processes. A function often used in reliability analysis is the *(negative) exponential distribution*. The (negative) exponential distribution is often used to describe the failure and repair processes of components. The exponential distribution is defined by the PDF and CDF [1]:

$$f_X(x) = \begin{cases} \lambda e^{-\lambda x} & x \geq 0 \\ 0 & \text{otherwise} \end{cases} \tag{A.14}$$

$$F_X(x) = \begin{cases} 1 - e^{-\lambda x} & x \geq 0 \\ 0 & \text{otherwise} \end{cases} \tag{A.15}$$

where:
λ = rate parameter ($\lambda > 0$)

Another common distribution is the *normal distribution* or *Gaussian distribution*:

$$f_X(x) = \frac{1}{\sqrt{2\pi\sigma^2}} e^{-\frac{(x-\mu)^2}{2\sigma^2}} \tag{A.16}$$

where:
μ = mean
σ = standard deviation

and the *Weibull distribution*:

$$f_X(x) = \frac{c}{\alpha} \left(\frac{x}{\alpha}\right)^{c-1} e^{-\left(\frac{x}{\alpha}\right)^c} \tag{A.17}$$

where:
c = shape parameter ($c \geq 0$)
α = scale parameter ($\alpha \geq 0$)

A.2 Statistics

For a population of outcomes of a probabilistic experiment, the *sample mean* is:

$$\bar{X} = \frac{1}{N} \sum_{i=1}^{N} X_i \tag{A.18}$$

where:
$\bar{X}$ = sample mean
N = number of outcomes
X_i = outcome i

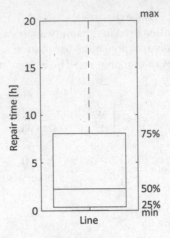

Fig. A.3 Boxplot example of repair times of overhead lines

Table A.1 Relation between boxplot and exponential distribution

Boxplot value	Relation with exponential distribution
Maximum	∞
Third quartile	$1.39 \cdot 1/\lambda$
Median	$0.69 \cdot 1/\lambda$
First quartile	$0.29 \cdot 1/\lambda$
Minimum	0

For an increasing number of observed outcomes, the sample mean will approach the expected value of the probabilistic process.

A *boxplot* is a way to present statistical data. It shows the *minimum* of the population, the first *quartile* (25% of the population is smaller than this value), the *median* (50% of the population is smaller than this value), the third *quartile* (75% of the population is smaller than this value), and the *maximum*. *Outliers* are sometimes excluded from the boxplot. Figure A.3 shows an example boxplot. The minimum is indicated by the lowest black line (on the axis line), the first quartile is shown by the lowest line of the box, the median is the line in the middle of the box, the third quartile is the upper line of the box, and the dashed line goes up to the maximum.

In Table A.1, the relationship between the boxplot and the exponential distribution is shown. It is typical for the exponential distribution that most of the outcomes are smaller than the average. The boxplot values can be calculated as follows:

$$e^{-\lambda t} = 0.25 \rightarrow t = 1.39\frac{1}{\lambda} \tag{A.19}$$

$$e^{-\lambda t} = 0.50 \rightarrow t = 0.69\frac{1}{\lambda} \tag{A.20}$$

$$e^{-\lambda t} = 0.75 \rightarrow t = 0.29\frac{1}{\lambda} \tag{A.21}$$

The average is located somewhere between the median (69% times mean) and the third quartile (139% times mean), and at the average lifetime, it is expected that 63% of the components have failed already:

$$t = \frac{1}{\lambda} \rightarrow F\left(\frac{1}{\lambda}\right) = 1 - e^{-\lambda \cdot \frac{1}{\lambda}} = 1 - \frac{1}{e} = 1 - 0.37 = 0.63 \tag{A.22}$$

Reference

1. Yates RD, Goodman DJ (2005) Probability and stochastic processes. Wiley, Amsterdam

Appendix B
Load Flow Calculations

Load flow (or power flow) calculations are used within most of the reliability analyses of power systems. While in some studies, a simple connectivity analysis is sufficient, in other studies a full AC load flow is required. This text is meant as a (very) brief introduction to load flow calculations. The interested reader is referred to textbooks like [1].

B.1 Connectivity Study

In some studies, a *connectivity study* is sufficient. In a connectivity study, it is analyzed whether certain points in the network are still connected to each other [2]. It is applied to systems where each component could take the complete system load. An example is the reliability analysis of substations, as illustrated in Fig. B.1. In case (a), points a, b, and c are all connected to each other. In case (b), point b is not connected anymore to points a and c, while these last two points are still connected to each other.

B.2 Graph Flow

In some other reliability studies, *graph flow analysis* can be applied. Actually, the application of graph flow analysis is limited to some special cases. Figure B.2 shows an example graph. As can be seen, point a is connected to point b by a network, and the capacity of each connection is as indicated in the figure.

There are several algorithms to find the maximum graph flow. One solution is illustrated in Fig. B.3. The algorithm is straightforward: Find one path through the network, find the maximum flow through this path, and reduce the connection capacities by this maximum path flow. Repeat this until points a and b are no longer connected. The maximum graph flow is the sum of the maximum path flows. As illustrated

© Springer Nature Switzerland AG 2020
B. W. Tuinema et al., *Probabilistic Reliability Analysis of Power Systems*,
https://doi.org/10.1007/978-3-030-43498-4

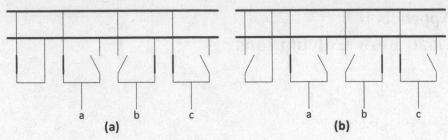

Fig. B.1 Substation connectivity study

Fig. B.2 Graph flow analysis

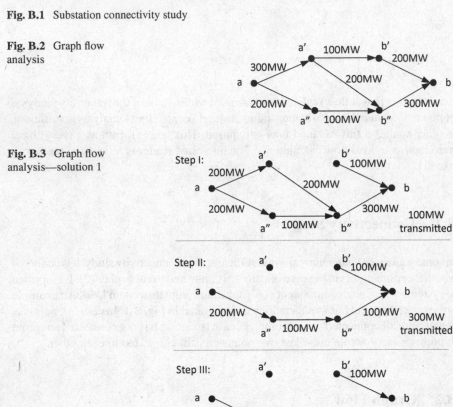

Fig. B.3 Graph flow analysis—solution 1

in Fig. B.3, this is done in three steps. First path a-a′-b′-b is considered, then path a-a′-b″-b, and finally path a-a″-b″-b.

Another solution is shown in Fig. B.4. According to the *maximum-flow, minimum-cut theorem* [3], 'the maximum amount of flow passing from the source to the sink is equal to the minimum capacity that, when removed in a specific way from the network, causes the situation that no flow can pass from the source to the sink.' As

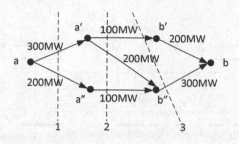

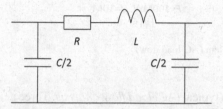

Fig. B.4 Graph flow analysis—solution 2

Fig. B.5 Model of a transmission line

shown in Fig. B.4, the graph can be cut into two pieces (disconnecting points a and b) in several ways. For example, the figure shows cuts 1, 2, and 3. It can easily be seen that the maximum flows through these cuts are 500MW, 400MW, and 400MW, respectively. The maximum graph flow is the minimum of these values: 400MW. Other possible cuts of the graph in the figure have larger maximum flows.

As mentioned before, the application of graph flow analysis in power system reliability analysis is limited to some special studies. It is clear that in these studies, there must be one source node and one sink node connected by a network. Furthermore, it is assumed that the flow through the network is directed such that every component is maximally loaded to its capacity and not more. This is seldom the case in real power systems. But in some cases, like a specific connection topology of offshore wind farms [4], graph flow can successfully be applied.

B.3 AC Load Flow

The actual load flow in a power system depends on various parameters like the impedance/admittance of network components and active/reactive power of the generators and loads. Generally, transmission lines are modeled as shown in Fig. B.5, where R represents the line resistance, L the line inductance, and C the line capacitance. Loads are typically represented by their active/reactive power consumption, while for generators mostly the active power production and the voltage amplitude are given.

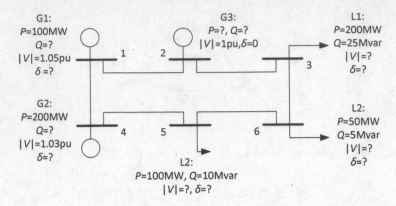

Fig. B.6 Load flow problem (AC load flow)

Figure B.6 shows a typical *(AC) load flow analysis*. Three generators (G1/G2/G3) and three loads (L1/L2/L3) are shown in the network. For the loads, the active/reactive power consumption is shown. For generators G1 and G2, the active power production and the voltage amplitude are given. The voltage is measured in pu (per unit), where 1 pu is the standard voltage of the network (e.g. 380 kV).

In a load flow problem, we would like to know the following information for each system bus: the active power consumption, the reactive power consumption, the voltage amplitude, and the voltage angle. Remember that the sinusoidal (three-phase) voltage can be represented by the voltage amplitude and the voltage angle [1]. Figure B.6 shows the known and unknown parameters for each system bus.

As can be seen, generator G3 is a special generator in the network. For this generator, the voltage amplitude and the voltage angle are given while the active/reactive power productions are unknown. The bus to which this generator is connected (bus 2) is called the swing bus (or slack bus) and has two main functions in the network. First, the swing bus compensates for mismatches in the total active/reactive power consumption and production in the network. These mismatches can be caused by network losses, but can as well be caused by uncertainties or errors in the active/reactive power of the other system buses. Second, the swing bus is the reference point for the voltage amplitude and voltage angle in the network.

An AC load flow problem can be solved by several solution techniques [1]. Generally, this is a complicated and iterative matrix solution process.

B.4 DC Load Flow

Because a full AC load flow is an iterative matrix solution, it can take much computation time when studying a multitude of load flows. Therefore, a simplification of the AC load flow exists in the form of the *DC load flow*.

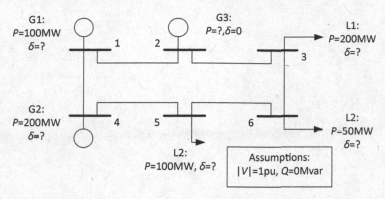

Fig. B.7 Load flow problem (DC load flow)

For a DC load flow problem, several assumptions are made. First, there are no losses in the network. This means that in Fig. B.5, R is neglected. Sometimes, C is neglected as well. Furthermore, there is no reactive power consumption/production in the network and the voltage amplitude is assumed 1 pu at all system buses.

With these assumptions, the load flow problem of Fig. B.6 reduces to the network in Fig. B.7. Because the DC load flow solution consists of only one matrix solution, the computational effort is reduced significantly compared to the full AC load flow, but at the cost of less accurate results.

References

1. Grainger JJ, Stevenson WD (jr) (1994) Power system analysis. McGraw-Hill International Editions
2. Li W (2005) Risk assessment of power systems - models, methods, and applications. Wiley Interscience - IEEE Press, Canada
3. Wikipedia (2015) Max-flow min-cut theorem. https://en.wikipedia.org/wiki/Max-flow_min-cut_theorem
4. Getreuer RE, Tuinema BW, Rueda JL, Meijden van der MAMM (2016) Multi-parameter approach for the selection of preferred offshore power grids for wind energy. In: IEEE international energy conference, EnergyCon2016, Leuven, Belgium

Appendix C
Reliability Indicators

For different parts of the power system, different sets of reliability indicators exist. Some of the most used reliability indicators are listed in this appendix [1–5].

Reliability indicators for the generation system:

- *LOLP*: *Loss Of Load Probability* [-]
  The probability that the demanded power cannot be supplied (partially or completely) by the generation system. The LOLP is often determined based on a per-hour study for a studied time period (usually a year);
- *LOLE*: *Loss Of Load Expectation* [h/y]
  Expected amount of time per period that the demanded power cannot be supplied (partially or completely) by the generation system;
- *LOEE*: *Loss Of Energy Expectation* [MWh/y]
  Total amount of energy that is expected not to be supplied during a given time period (usually on a yearly basis) because of failures of the generation system;
- *Capacity Credit*: [MW]
  The amount of conventional capacity that can be replaced by a (variable) energy source, while keeping the same level of reliability;
- *ESWE*: *Expected Surplus Wind Energy* [MWh/y]
  Expected amount of wind energy that cannot be absorbed by the system;
- *ELCC*: *Effective Load Carrying Capability* [MW]
  The amount of load that can be served by a (variable) energy source, while keeping the reliability of the system at the same level;
- *LOCP*: *Loss Of Capacity Probability* [-]
  Probability of having a certain generation capacity outage.

Security-of-supply-related indicators:

- *PLC*: *Probability of Load Curtailment* [-]
  The probability that the demanded load cannot be supplied (partially or completely). The PLC is often determined based on a set of considered contingencies (single contingencies as well as higher-order contingencies) and a studied time period (usually on a yearly basis);

B. W. Tuinema et al., *Probabilistic Reliability Analysis of Power Systems*,
https://doi.org/10.1007/978-3-030-43498-4

- *EENS*: *Expected Energy Not Supplied* [MWh]
  Total amount of energy that is expected not to be supplied during a given time period (usually on a yearly basis) due to supply interruptions;
- *SAIDI*: *System Average Interruption Duration Index* [min]
  Average outage duration of each customer during a given time period (usually on a yearly basis);
- *SAIFI*: *System Average Interruption Frequency Index* [-]
  Average number of interruptions per customer during a given time period (usually on a yearly basis);
- *CAIDI*: *Customer Average Interruption Duration Index* [min]
  Average interruption duration per interrupted customer during a given time period (usually on a yearly basis);
- *ASAI*: *Average Service Availability Index* [-]
  Ratio between the available service duration and the demanded service duration.

  Additional reliability indicators:

- *Probability of Overload* [-]
  The probability that one or more connections in the network are overloaded during a given time period (usually on a yearly basis);
- *Probability of Generation Redispatch* [-]
  The probability that generation redispatch is applied during a given time period (usually on a yearly basis);
- *Expected Redispatch Costs* [currency]
  Expected costs of generation redispatch during a given time period (usually on a yearly basis);
- *Probability of the Alert State* [-]
  The probability that the power system is in the alert state during a given time period (usually on a yearly basis).

References

1. Billinton R, Allan RN (1996) Reliability evaluation of power systems, 2nd edn. Plenum Press, New York
2. Kandalepa N, Tuinema BW, Rueda JL, Meijden van der MAMM (2015) Reliability modeling of transmission networks: an explanatory study on further EHV underground cabling in the Netherlands. MSc Thesis, Department of Electrical Engineering, Delft University of Technology, Delft, the Netherlands
3. Li W (2011) Probabilistic transmission system planning. Wiley Interscience - IEEE Press, Canada
4. Lemstrom B et al. (2008) Effects of increasing wind power penetration on power flows in European grids. Tradewind
5. Wu L, Park J, Choi J, El-Kaib AA, Shahidehpour M, Billinton R (2009) Probabilistic reliability evaluation of power systems including wind turbine generators using a simplified multi-state model: a case study. In: IEEE power and energy society general meeting 2009 (PES'09)

Appendix D
Solutions

Problems of Chap. 1

1.1 Learning Objectives
The learning objectives can be found at the beginning of the chapter.

1.2 Main Concepts
The description of these concepts can be found in the text of the chapter.

1.3 System Reliability
Reliability is the ability of the system to fulfill its function (under certain conditions and for a certain period). The main function of the (traditional) power system is to supply electricity to the load.

1.4 Liberalization Electricity Market
In the liberalization of the power system, the production and transport of electricity were divided over different parties. The liberalization of the power system therefore led to various new actors: producers, consumers, system operators, electricity traders, service providers, etc. The main function of the power system changed to: connecting customers (i.e. consumers/producers/etc.) and enable the trade in electricity. This development also changed the concept of power system reliability. If a customer (i.e. consumer/producer/etc.) is interrupted or hindered when trading in electricity, the power system can be regarded as unreliable.

1.5 Business Values
The six business values are: secure supply, engage stakeholders, safety, financial, environment, and compliance. Security of supply is directly related to reliability as it reflects how often consumers are interrupted. However, blackouts can also lead to financial risks (e.g. compensation costs), engage stakeholders (e.g. customer claims), compliance (e.g. the TSO did not act according to the law), such that these business values are also related to reliability. Safety (e.g. the number of injuries after an accident) and environment (e.g. environmental damage caused by an accident) might be less directly related to reliability, but this is open for discussion.

© Springer Nature Switzerland AG 2020
B. W. Tuinema et al., *Probabilistic Reliability Analysis of Power Systems*,
https://doi.org/10.1007/978-3-030-43498-4

1.6 Probabilistic and Deterministic Analysis

In deterministic analysis, the effects of certain (often: worst-case) scenarios are studied to calculate results for each scenario, while in probabilistic analysis a range of possible scenarios is included to calculate one result. Deterministic analysis does not include the probabilities of the scenarios and component failures while probabilistic analysis does. For example, in deterministic analysis it can be studied whether a component is able to withstand the 'standard' lightning pulse or whether a network is $N - 1$ redundant. This leads to a single result: OK or not. In probabilistic analysis, the effects of a range of possible lightning pulses or the impact of higher-order failures on the network reliability can be studied. Possible criteria are $N - 1$ redundancy (deterministic) or a probability of load curtailment smaller than x (probabilistic).

1.7 TSO Activities

Examples of TSO activities are shown in Fig. 1.7 (Chap. 1).

1.8 Interaction of TSO Activities

A good example is the discussion about redundancy in offshore networks. Should these offshore networks be $N - 1$ redundant like the onshore transmission network? But then, the costs increase much and the offshore network might become uneconomical. If the offshore network is not $N - 1$ redundant, failures of the offshore network can result in large (wind) capacity outages. Then, it is likely that preventive and corrective control actions must be applied more often during system operation.

1.9 Control Actions

Some corrective/preventive control actions are: Use phase-shifting transformers (PSTs) to change the load flow, perform switching actions or network reconfigurations, perform generation redispatch, and cancel maintenance, wind curtailment and load curtailment. These are in order of preference, but this can differ from case to case.

Problems of Chap. 2

2.1 Learning Objectives

The learning objectives can be found at the beginning of the chapter.

2.2 Main Concepts

The description of these concepts can be found in the text of the chapter.

2.3 Failure Statistics

To obtain more failure statistics, one can:

- Collect failure statistics for a longer period: This will of course take longer, but also time effects like the introduction of improved component designs in recent years might be neglected if older failure statistics are included as well.
- Combine the failure statistics of different voltage levels: In this way, the effects of other component designs for different voltage levels are neglected, and also a larger amount of statistics of lower voltage levels might dominate the results.

- Combine the failure statistics with failure statistics of other power systems: This is possible if the power systems are comparable (in design and environment); for example, Dutch failure statistics probably cannot be combined with failure statistics from Spain because of the different climates.

2.4 Failure Frequencies of EHV/HV

HV overhead lines have a larger average failure frequency than EHV overhead lines. There is no overlap in the confidence intervals, such that the difference is significant. For transformers, it is the other way round: EHV transformers have a larger failure frequency than HV transformers. For underground cables, the confidence interval is very large for EHV cables. There are too few failure statistics to calculate an accurate failure frequency. For HV cables, there are enough failure statistics and the confidence interval is reasonable.

2.5 Repair Times

There are several challenges of defining accurate repair times. The amount of available failure statistics can be a problem for new component types (like EHV underground cables). But also different modeling methods (e.g. averages, boxplots, exponentials) can result in different values for the repair time. Extremely long repairs tend to be dominant when calculating the average repair time. Also, the definition of 'repair' (e.g. actual component repair, time between failure and back in operation, time to emergency solution) can cause some confusion.

2.6 Failure Progress

Some examples are:

- Failure causes: weather, manufacturing defects, aging/wear, third-party damage;
- Failure mechanisms: treeing (in insulation material) and mechanical/electrical wear;
- Failure modes: broken overhead line, exploded current transformer, low oil level;
- Faults: line–ground fault, line–line fault, 3-phase (symmetrical) fault, 3-phase ground (symmetrical) fault, 'short circuit,' overvoltage.

2.7 Failure Progress Examples

One example is: impurity in insulation material, treeing, insulation breakdown, short circuit. Another example is: cold weather and storms, line dancing/galloping (mechanical wear), broken isolator, line–ground fault.

2.8 Failure Causes

A too simple answer would be: to build stronger components that last longer. The graphs of the failure causes show that many component failures are caused by external causes. These can be weather effects, but many failures are caused by third-party damage. Third-party damage can be prevented by creating a certain awareness, e.g. by prevention campaigns. For example, before excavation works are performed, first a detailed map of the underground situation must be considered, which is required by law. But also warning signs that indicate the location of a cable can be helpful.

Some underground cables are even installed within concrete tubes to prevent third-party damage. As most overhead line failures are caused by external causes and hoisting works as well, it seems that it can be improved by making people aware of the possible danger when working in the vicinity of an overhead line.

2.9 Correlation and Dependency

In the case of dependency, there is a causal relationship. In the case of correlation, there is a common pattern but a causal relationship does not have to be present. If underground cable failures occur more often during the day, the failure causes have to be studied in more detail. Dependency can be possible if many cable failures are caused by excavation works, which are mostly performed during daytime.

2.10 Failure Modes

When circuit breakers and disconnectors fail, they often fail when switching is required. For example, they do not open or close when needed. Considering this, it might be more logical to express the failure frequency of these components in the unit [per switching action] rather than [per year]. The first provides a useful conditional probability for reliability analysis. The second might however be useful in asset management, as it could be used as indication for how often these components need maintenance.

2.11 Protection System

Although protection systems are used to switch off failed components to prevent worse, protection system can fail themselves as well. Spontaneous protection system switching is one example. But also non-selective switching when the protection system is operated is an example. In this case, more components can be switched off than decided. Refusing circuit breakers are an example as well.

2.12 Largest Blackouts in the Netherlands

For Table 2.8:

1. It seems to be neither cascading nor common cause, and the (human) misinterpretation and the fact that it happened during maintenance works (and the protection system could not switch selectively) are complicating factors.
2. Typical common cause: Two circuits out because of a helicopter, where the complicating factor was the long repair time (because of high water in the river).
3. Protection system malfunctioning is a complicating factor here, and the overloading of the third transformer is a cascading failure.
4. This seems to be a cascading failure: Initially, only one transformer failed, and this caused fire such that the complete substation was switched off.
5. This seems to be a cascading failure as well: Because initially there was only a short circuit in the circuit breaker, it initiated a fire and the outage of more components.
6. Neither: Complicating factor is that the failure occurred during maintenance.

7. Cascading: Initially, only the voltage transformer failed, but because the protection system did not function properly, more components were switched off. The fact that the protection system did not function properly can be seen as a complicated factor.

8. Cascading: The failure of the first transformer caused an overloading of the third, and complicating factor is that it happened during maintenance of transformer two.

9. This is not really clear: Two transformers have failed, and the description speaks about a complex failure of multiple components, but the description is too short to give clear insight into this case.

10. Typical common cause: Two circuits failed simultaneously because of a tall boat.

In general, the distinction between common cause and cascading is not always clear in reality. Also, this exercise shows that large blackouts are mostly the results of a complex combination of failures and complicating factors.

2.13 Largest Blackouts in the World
The description depends on the blackout that is chosen. Carefully think about whether you describe the concepts (failure cause, failure mode, dependency and correlation, other factors, etc.) well for this specific blackout.

Problems of Chap. 3

3.1 Learning Objectives
The learning objectives can be found at the beginning of the chapter.

3.2 Main Concepts
The definitions can be found in the text of the chapter.

3.3 Reliability Functions of the Component
This exercise is comparable to Example 3.1 in the chapter. We first start with the reliability function ($R(t)$). At time $t = 0$, all components are still working, so the reliability is $R(0) = 1$. Then, after 2 years, one component fails. As 90% of the components are still working then, the reliability function becomes $R(2) = 0.9$. After 4 years, a second component fails. As 80% of the components are still working then, the reliability function becomes $R(4) = 0.8$. This continues for the other components, until all components have failed after 36 years ($R(36) = 0$). Table D.1 shows the values of the reliability function. The reliability function is drawn in Fig. D.1.

The next step is the calculation of the unreliability function. As the unreliability function is $F(t) = 1 - R(t)$, it can be calculated directly from the reliability function. The values are shown in Table D.1, and the unreliability function is drawn in Fig. D.1 as well.

The failure density distribution function then is the derivative of the unreliability function. Therefore, we can calculate that it is:

Table D.1 Reliability functions

t (y)	$R(t)$ (-)	$F(t)$ (-)	$f(t)$ (/y)	$h(t)$ (/y)
0	1	0	1/20	0.050
2	0.9	0.1	1/20	0.056
4	0.8	0.2	1/20	0.063
6	0.7	0.3	1/30	0.048
9	0.6	0.4	1/30	0.056
12	0.5	0.5	1/40	0.050
16	0.4	0.6	1/40	0.063
20	0.3	0.7	1/40	0.083
24	0.2	0.8	1/60	0.083
30	0.1	0.9	1/60	0.167
36	0	1	0	–

$$f(t) = \frac{1}{20} / y \qquad \text{for } 0 \leq t < 6,$$

$$f(t) = \frac{1}{30} / y \qquad \text{for } 6 \leq t < 12,$$

$$f(t) = \frac{1}{40} / y \qquad \text{for } 12 \leq t < 24,$$

$$f(t) = \frac{1}{60} / y \qquad \text{for } 24 \leq t < 36, \text{ and}$$

$$f(t) = 0 / y \qquad \text{for } 36 \leq t. \tag{D.1}$$

The values of the failure density distribution function are shown in Table D.1, and the failure density distribution function is drawn in Fig. D.1.

The hazard rate can be calculated from the failure density function by pointwise dividing the values of the first by the values of the latter. The values of the hazard rate then become as shown in Table D.1. The hazard rate is drawn in Graph D.1. At $t \rightarrow 36$, the hazard rate increases to infinity, and at $t \geq 36$, the hazard rate is undefined as it is $0/0$.

3.4 Reliability Functions

There are four reliability functions:

- $F(t)$: Unreliability function or failure distribution. This is the probability of finding an originally healthy component in a failed state after time t.
- $R(t)$: Reliability function. This is the probability of finding a healthy component in a healthy state after time t:

$$R(t) = 1 - F(t) \tag{D.2}$$

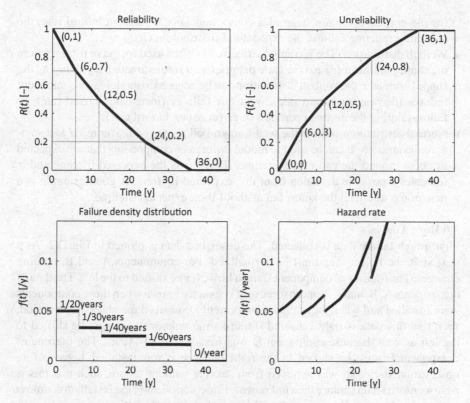

Fig. D.1 Reliability functions for the considered component

- $f(t)$: Failure density distribution. This is the rate at which a component fails at time t. The failure density distribution is the derivative of the failure distribution:

$$f(t) = \frac{dF(t)}{dt}$$
(D.3)

- $h(t)$: Hazard rate. This is the rate at which a component fails at time t, given that it is healthy until time t:

$$h(t) = \frac{f(t)}{R(t)}$$
(D.4)

3.5 Probability Distributions

Three probability distributions that are often used are the (negative) exponential distribution, the Weibull distribution, and the normal (or Gaussian) distribution.

- Exponential distribution: This distribution is very often used, where it is assumed that the hazard rate is a constant, the failure rate λ. This is a reasonable assumption for the normal operation stage of a component. For a constant hazard rate, the reliability function follows the exponential distribution ($R(t) = e^{-\lambda t}$).

- Weibull distribution: The Weibull distribution is often used for curve fitting, where the shape parameter (c) and the scale parameter (α) are estimated. By changing the shape parameter, the Weibull distribution can be adapted to model early component failures (decreasing hazard rate), wear-out failures (increasing hazard rate), or failures during the normal operation stage (constant hazard rate).
- Normal distribution: This is the well-known bell shape, with average μ and standard deviation σ. It can be used to model failures of components that are expected to occur around the end of the lifetime. Then, μ is the expected lifetime and σ indicates a possible deviation from this expected lifetime. A good example is a new office where all the lamps fail at about their expected lifetime.

3.6 Data Analysis

First, rough failure data is collected. The described data is plotted in Fig. D.2. As a next step, the failure data must be normalized. For components A and B, nothing changes. The lifetime of component C must however be shifted to the left. The data of components A, B, and C is uncensored as it is exactly known when these components were installed and when they failed. Component D survived the observation period, such that this data is right censored (failure time unknown). It must be shifted to the left as well, because component D was installed after 4 years. The lifetime of component E must be shifted to the right because it was installed 1 year before observation started. It was removed from service without failure, such that this is right-censored data (failure time unknown). For component F, the installation time is unknown, which is left-censored data. Because the exact installation time is unknown, the lifetime cannot be shifted to the right, but the circle indicated that this component lived longer than 4 years.

The final step is to sort the data, as shown in Fig. D.2. Then, the failure distribution can be estimated, for example, by using computer software. This computer software can also take into account a different order of the component lifetimes caused by the

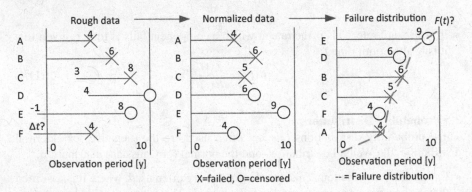

Fig. D.2 Failure data analysis

censored data. For example, a component with an unknown failure time might have lived longer than a component with a longer observed lifetime and a known failure time.

3.7 Bathtub Curve
See Figs. 3.4 and 3.5 in Chap. 3.

3.8 Negative Exponential Distribution
(Negative) exponential distribution:

- The (negative) exponential distribution is a way to model the reliability behavior of components by a probability distribution.
- The main assumption is that the hazard rate is a constant, namely the failure rate ($h(t) = \lambda$).
- The reliability function follows the (negative) exponential distribution:

$$R(t) = e^{-\lambda t} \tag{D.5}$$

- The other functions are:

$$F(t) = 1 - R(t) = 1 - e^{-\lambda t} \tag{D.6}$$

$$f(t) = \frac{dF(t)}{dt} = \lambda e^{-\lambda t} \tag{D.7}$$

$$h(t) = \frac{f(t)}{R(t)} = \frac{\lambda e^{-\lambda t}}{e^{-\lambda t}} = \lambda \tag{D.8}$$

- The graphs are shown in Fig. 3.6, Chap. 3.

3.9 Weibull Analysis
In Weibull analysis, a Weibull distribution is fit to some failure data. By varying the shape parameter (c) of the Weibull distribution, the Weibull distribution can model an increasing, decreasing, or constant hazard rate. The scale parameter (α) is used for scaling of the distribution. This is illustrated in Fig. D.3. By performing Weibull analysis, it is possible to estimate the shape parameter, such that it is known whether the hazard rate is increasing, decreasing, or constant. Then, it is also known whether the specific component is in its infant stage, normal operation stage, or wear-out stage. An example for Weibull analysis of cable terminations is illustrated in Fig. D.4.

3.10 Component Life Cycle
The life cycle is shown in Fig. D.5:

$$f = \frac{1}{MTBF} = \frac{1}{MTTF + MTTR} \tag{D.9}$$

$$\lambda = \frac{1}{MTTF} = \frac{1}{MTBF - MTTR} = \frac{1}{\frac{1}{f} - r} \tag{D.10}$$

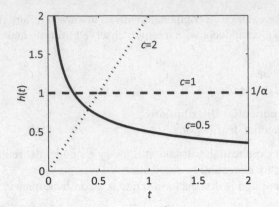

Fig. D.3 Weibull hazard rate

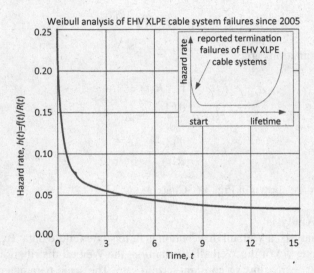

Fig. D.4 Weibull example

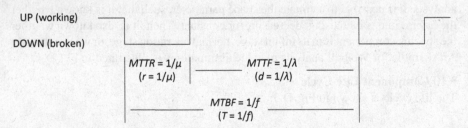

Fig. D.5 Life cycle of a component

3.11 Failure Rate Transformer

$$f_{tr} = 0.05 \, /y \tag{D.11}$$

$$r_{tr} = \frac{1460}{8760} = 0.167 \, y \tag{D.12}$$

$$U_{tr} = \frac{MTTR_{tr}}{MTBF_{tr}} = f_{tr} r_{tr} = 0.05 \cdot 0.167 = 8.33 \cdot 10^{-3} \tag{D.13}$$

For this (dimensionless) unavailability, it is expected that the transformer is $8760 \cdot 8.33 \cdot 10^{-3} = 73 \, h/y$ unavailable.

$$\lambda_{tr} = \frac{1}{MTTF_{tr}} = \frac{1}{MTBF_{tr} - MTTR_{tr}} = \frac{1}{\frac{1}{f_{tr}} - r_{tr}} = \frac{1}{\frac{1}{0.05} - 0.167} = 0.0504 \, /y \tag{D.14}$$

3.12 Repair Time Overhead Line

$$U_{ohl} = \frac{0.1}{8760} \tag{D.15}$$

$$f_{ohl} = 0.1 \, /y \tag{D.16}$$

$$r_{ohl} = MTTR_{ohl} = \frac{MTTR_{ohl} MTBF_{ohl}}{MTBF_{ohl}} = MTBF_{ohl} \frac{MTTR_{ohl}}{MTBF_{ohl}}$$
$$= \frac{U_{ohl}}{f_{ohl}} = \frac{0.1}{8760 \cdot 0.1} = 1.14 \cdot 10^{-4} \, y \tag{D.17}$$

This equals $8760 \cdot 1.14 \cdot 10^{-4} = 1 \, h$.

3.13 Failure Frequencies and Repair Times
In the figures of the failure frequencies (Fig. 2.1), it can be clearly seen that the confidence intervals are much larger if the amount of available failure statistics is small (compare this with Example 3.10 in Chap. 3). The table with repair times (Table 2.3) shows that different modeling approaches can lead to significantly different repair times (compare this with Example 3.5 in Chap. 3).

3.14 Markov Model Offshore Cable
The failure frequency and repair time are given, but the failure rate and repair rate are needed for the two-state Markov model.

$$r_{oc} = 730/8760 = 0.0833 \, y \tag{D.18}$$

$$\mu = 1/r_{oc} = 1/0.0833 = 12 \, /y \tag{D.19}$$

Fig. D.6 Markov model
offshore cable

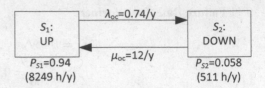

$$\lambda_{oc} = \frac{1}{\frac{1}{f_{oc}} - r_{oc}} = \frac{1}{\frac{1}{0.70} - 0.0833} = 0.743\,/y \tag{D.20}$$

$$P_{S_1} = A_{oc} = \frac{MTTF_{oc}}{MTBF_{oc}} = \frac{f_{oc}}{\mu_{oc}} = \frac{0.70}{0.74} = 0.942\ (=8249\,h/y) \tag{D.21}$$

$$P_{S_2} = U_{oc} = \frac{MTTR_{oc}}{MTBF_{oc}} = r_{oc}f_{oc} = 0.0833 \cdot 0.70 = 0.058\ (=511\,h/y) \tag{D.22}$$

The Markov model is shown in Fig. D.6.

To solve the Markov model in equilibrium, the frequencies of the state transitions (entering a state and leaving a state) must be equal (i.e. $f_{S_1 \to S_2} = f_{S_2 \to S_1}$, so: $P_{S_1}\lambda_{oc} = P_{S_2}\mu_{oc}$). Furthermore, the sum of the state probabilities must (always) equal 1 (i.e. $P_{S_1} + P_{S_2} = 1$).

3.15 Markov Model Generator

The failure frequency and repair time are given, but the failure rate and repair rate are needed for the two-state Markov model.

$$r_{gen} = 48/8760 = 5.48 \cdot 10^{-3}\,y \tag{D.23}$$

$$\mu = 1/r_{gen} = \frac{1}{5.48 \cdot 10^{-3}} = 182.5\,/y \tag{D.24}$$

$$\lambda_{gen} = \frac{1}{\frac{1}{f_{gen}} - r_{gen}} = \frac{1}{\frac{1}{5} - 5.48 \cdot 10^{-3}} = 5.14\,/y \tag{D.25}$$

$$P_{S_1} = A_{gen} = \frac{MTTF_{gen}}{MTBF_{gen}} = \frac{f_{gen}}{\mu_{gen}} = \frac{5}{5.14} = 0.973\ (=8520\,/y) \tag{D.26}$$

$$P_{S_2} = U_{gen} = \frac{MTTR_{gen}}{MTBF_{gen}} = r_{gen}f_{gen} = 5.48 \cdot 10^{-3} \cdot 5 = 0.027\ (=240\,/y) \tag{D.27}$$

The Markov model is shown in Fig. D.7. The generator is (on average) 240 h/y unavailable.

3.16 Stress-Strength Model

The stress-strength model is shown in Fig. D.8. It is assumed that both the strength of a component and the stress on the component can be described by probability

Fig. D.7 Markov model generator

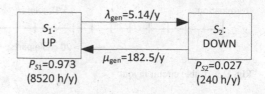

Fig. D.8 Stress-strength model

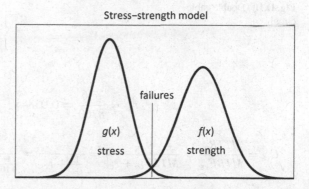

distributions, in this case two normal (Gaussian) distributions. If the stress on the component is larger than the strength of the component, the component fails. The area where component failures occur is indicated in the graph as well.

Problems of Chap. 4

4.1 Learning Objectives
The learning objectives can be found at the beginning of the chapter.

4.2 Main Concepts
The definitions can be found in the text of the chapter.

4.3 Unavailability Cable Circuit
The layout of the cable circuit is shown in Fig. D.9.

As it is a series connection, the failure rates can be added. Please note that the length of the circuit is 20 km and the failure rate of the cable parts is given in [/cctkmyr] (per circuit-kilometer year). In this example, the cable parts happen to be exactly 1 km each.

$$\lambda_{cs} = 20\lambda_c + 19\lambda_j + 2\lambda_t = 0.10/y \tag{D.28}$$

For calculating the unavailability of this circuit, it is important to know that all components have the same repair time. Example 4.8 in the reader namely shows that if the repair times are the same, all possible component failures can be combined leading to a two-state Markov model. If the different parts of the cable circuit have different repair times, components must be grouped into groups with the same repair time, leading to a multiple state Markov model. If all components have the same repair time:

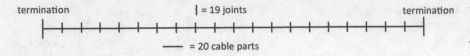

Fig. D.9 Cable circuit layout

Fig. D.10 Double cable circuit

$$r_{cs} = \frac{336}{8760} = 0.0384\,\text{y} \tag{D.29}$$

$$\dot{U}_{cs} = \frac{MTTR_{cs}}{MTBF_{cs}} = \frac{r_{cs}}{MTTF_{cs} + r_{cs}} = \frac{r_{cs}}{\frac{1}{\lambda_{cs}} + r_{cs}} = \frac{\frac{336}{8760}}{\frac{1}{0.10} + \frac{336}{8760}} = 0.0038 \tag{D.30}$$

This is (on average) $8760 \cdot 0.0038 = 33.5\,\text{h/y}$.

Then, the failure frequency becomes:

$$f_{cs} = \frac{1}{MTBF_{cs}} = \frac{1}{MTBF_{cs}} = \frac{1}{MTTF_{cs} + MTTR_{cs}}$$
$$= \frac{1}{\frac{1}{\lambda_{cs}} + MTTR_{cs}} = \frac{1}{\frac{1}{0.10} + \frac{336}{8760}} = 0.10\,/\text{y} \tag{D.31}$$

Or: $f_{cs} \approx \lambda_{cs}$.

Two circuits now form a double circuit, for which the reliability network of Fig. D.10 can be drawn.

The probabilities of having 0, 1, or 2 circuits available can be calculated as:

$$P_{full} = A_{cs}A_{cs} = (1 - U_{cs})^2 = (1 - 0.0038)^2 = 0.992\ (=8693\,\text{h/y}) \tag{D.32}$$

$$P_{half} = 2A_{cs}U_{cs} = 2(1 - U_{cs})U_{cs} = 2(1 - 0.0038)0.0038$$
$$= 7.61 \cdot 10^{-3}\ (=66.7\,\text{h/y}) \tag{D.33}$$

$$P_{zero} = U_{cs}U_{cs} = 0.0038^2 = 1.46 \cdot 10^{-5}\ (=0.13\,\text{h/y}) \tag{D.34}$$

4.4 Reliability Offshore Network

Because this particular network is $N - 1$ redundant, every component in the network is able to carry the full capacity of the wind farm. For example, if the transformer right below fails, all power is transported through the cable and transformer right above similar to the other components in the network. Because this network is $N - 1$ redundant, all components are designed for the total wind farm capacity and the main question is whether the offshore network is connected to the shore or not (i.e. a connectivity study; see also Appendix B). The network can then be solved using the techniques for series/parallel networks. If the network is not $N - 1$ redundant, a component failure might lead to a limited transport capacity. This network must then be solved using advanced techniques (e.g. fault tree analysis and state enumeration).

4.5 Reliability Network Cable

The underground cable connection is a series connection, such that failure rates can be added. Also, the availability of the connection is the product of the availabilities of the individual component. The Markov model in this example shows how failure states can be combined if the repair times are the same. Also, compare this with question 4.3.

4.6 Markov Model Generators

This question is quite similar to Examples 4.7 and 4.10 in the reader. The repair rate is:

$$\mu_g = \frac{1}{r_g} = \frac{1}{\frac{24}{8760}} = 365/y \tag{D.35}$$

The switching rate is:

$$\mu_{sw} = \frac{1}{r_{sw}} = \frac{1}{\frac{4}{8760}} = 2190/y \tag{D.36}$$

The Markov models are shown in Figs. D.11, D.12, D.13 and D.14. If the generator is not repaired during the switching time and the spare generator does not fail, the Markov model becomes as shown in option 4. This is also the normal progress in time through the Markov model.

If the failed generator can be repaired within the switching time, an extra state transition from state S_2 to state S_1 is drawn and the Markov model becomes as shown in option 3.

If the spare generator can fail as well then, an additional failure state (S_4) in which the spare generator has failed is drawn. In this state (both generators failed), it is possible that first the primary generator is repaired (transition to state S_3) or first the spare generator is repaired (transition to state S_5). Please notice that according to the Markov models, both generators are repaired simultaneously, but one repair will be completed first (and the repair of the other still continues). In option 1, the primary generator can also be repaired within the switching time, and in option 2, the primary generator cannot be repaired within the switching time.

4.7 Markov Model Variations

The Markov models are shown in Figs. D.15 and D.16. If both circuits are repaired simultaneously, where it is meant that the repair of both circuits is completed at the same time, the Markov model becomes as shown in case 1. There is now one state

Fig. D.11 Markov model
generators (option 1)

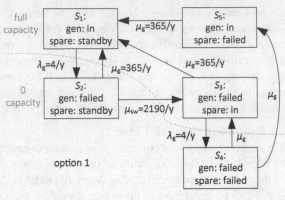

Fig. D.12 Markov model
generators (option 2)

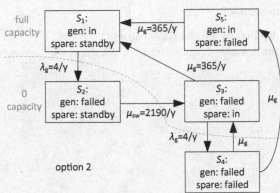

Fig. D.13 Markov model
generators (option 3)

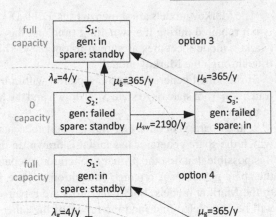

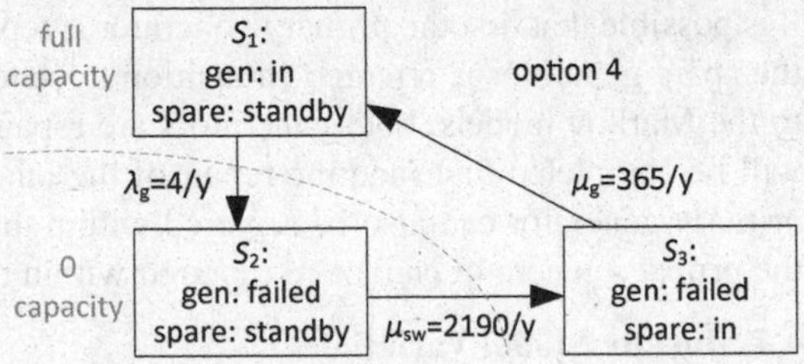

Fig. D.14 Markov model
generators (option 4)

Fig. D.15 Markov model overhead lines (Case 1)

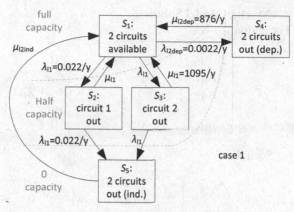

Fig. D.16 Markov model overhead lines (Case 2)

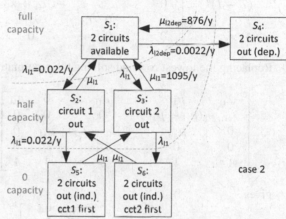

transition from state S_5 to state S_1. The state transitions from state S_5 to states S_2 and S_3 are removed.

If there is only one repair team that first repairs the circuit that failed first, the Markov model becomes as shown in case 2. State S_5 is now split into two states to make a distinction whether circuit 1 or 2 failed first. The state transitions from these state are now crossed (S_5 to S_3 and S_6 to S_2) such that the circuit which failed first is also repaired first.

Please notice that in Markov models it sometimes seems like a component repair starts over again. For example, if circuit 1 fails the system goes to state S_2 and the repair of circuit 1 starts. But if circuit 2 also fails, the system goes to state S_5 and it seems like the repair of circuit 1 starts over again (transition from state S_5 to state S_3). However, the memorylessness of the exponential distribution is used here, such that the expected time till completion of the repair does not depend on when the repair started.

4.8 Markov Model Solution

For option 1:

$$T = \begin{bmatrix} 1 - \lambda_{t1} & \lambda_{t1} & 0 & 0 & 0 \\ \mu_{t1} & 1 - (\mu_{t1} + \mu_{sw}) & \mu_{sw} & 0 & 0 \\ \mu_{t1} & 0 & 1 - (\mu_{t1} + \lambda_{t1}) & \lambda_{t1} & 0 \\ 0 & 0 & \mu_{t1} & 1 - 2\mu_{t1} & \mu_{t1} \\ \mu_{t1} & 0 & 0 & 0 & 1 - \mu_{t1} \end{bmatrix} \tag{D.37}$$

Such that:

$$(T - I)^{\text{tr}} = \begin{bmatrix} -\lambda_{t1} & \mu t1 & \mu t1 & 0 & \mu_{t1} \\ \lambda_{t1} & -(\mu_{t1} + \mu sw) & 0 & 0 & 0 \\ 0 & \mu_{sw} & -(\mu_{t1} + \lambda_{t1}) & \mu_{t1} & 0 \\ 0 & 0 & \lambda_{t1} & -2\mu_{t1} & 0 \\ 0 & 0 & 0 & \mu_{t1} & -\mu_{t1} \end{bmatrix} \tag{D.38}$$

Replacing the lowest row and solving the next matrix equation:

$$\begin{bmatrix} -\lambda_{t1} & \mu_{t1} & \mu_{t1} & 0 & \mu_{t1} \\ \lambda_{t1} & -(\mu_{t1} + \mu_{sw}) & 0 & 0 & 0 \\ 0 & \mu_{sw} & -(\mu_{t1} + \lambda_{t1}) & \mu_{t1} & 0 \\ 0 & 0 & \lambda_{t1} & -2\mu_{t1} & 0 \\ 1 & 1 & 1 & 1 & 1 \end{bmatrix} \begin{bmatrix} P_{S_1} \\ P_{S_2} \\ P_{S_3} \\ P_{S_4} \\ P_{S_5} \end{bmatrix} = \begin{bmatrix} 0 \\ 0 \\ 0 \\ 0 \\ 1 \end{bmatrix} \tag{D.39}$$

gives

$$P = \begin{bmatrix} 0.9994 \\ 2.61 \cdot 10^{-5} \\ 5.49 \cdot 10^{-4} \\ 1.58 \cdot 10^{-7} \\ 1.58 \cdot 10^{-7} \end{bmatrix}^{\text{tr}} \tag{D.40}$$

For option 2:

$$T = \begin{bmatrix} 1 - \lambda_{t1} & \lambda_{t1} & 0 & 0 & 0 \\ 0 & 1 - \mu_{sw} & \mu_{sw} & 0 & 0 \\ \mu_{t1} & 0 & 1 - (\mu_{t1} + \lambda_{t1}) & \lambda_{t1} & 0 \\ 0 & 0 & \mu_{t1} & 1 - 2\mu_{t1} & \mu_{t1} \\ \mu_{t1} & 0 & 0 & 0 & 1 - \mu_{t1} \end{bmatrix} \tag{D.41}$$

$$(T - I)^{\text{tr}} = \begin{bmatrix} -\lambda_{t1} & 0 & \mu t1 & 0 & \mu_{t1} \\ \lambda_{t1} & -\mu_{sw} & 0 & 0 & 0 \\ 0 & \mu_{sw} & -(\mu_{t1} + \lambda_{t1}) & \mu_{t1} & 0 \\ 0 & 0 & \lambda_{t1} & -2\mu_{t1} & 0 \\ 0 & 0 & 0 & \mu_{t1} & -\mu_{t1} \end{bmatrix} \tag{D.42}$$

$$\begin{bmatrix} -\lambda_{t1} & 0 & \mu t1 & 0 & \mu_{t1} \\ \lambda_{t1} & -\mu_{sw} & 0 & 0 & 0 \\ 0 & \mu_{sw} & -(\mu_{t1}+\lambda_{t1}) & \mu_{t1} & 0 \\ 0 & 0 & \lambda_{t1} & -2\mu_{t1} & 0 \\ 0 & 0 & 0 & \mu_{t1} & -\mu_{t1} \end{bmatrix} \begin{bmatrix} P_{S_1} \\ P_{S_2} \\ P_{S_3} \\ P_{S_4} \\ P_{S_5} \end{bmatrix} = \begin{bmatrix} 0 \\ 0 \\ 0 \\ 0 \\ 1 \end{bmatrix} \tag{D.43}$$

$$P = \begin{bmatrix} 0.99994 \\ 2.74 \cdot 10^{-5} \\ 5.75 \cdot 10^{-4} \\ 1.65 \cdot 10^{-7} \\ 1.65 \cdot 10^{-7} \end{bmatrix}^{tr} \tag{D.44}$$

For option 3:

$$T = \begin{bmatrix} 1-\lambda_{t1} & \lambda_{t1} & 0 \\ \mu_{t1} & 1-(\mu_{t1}+\mu_{sw}) & \mu_{sw} \\ \mu_{t1} & 0 & 1-\mu_{t1} \end{bmatrix} \tag{D.45}$$

$$(T-I)^{tr} = \begin{bmatrix} -\lambda_{t1} & \mu_{t1} & \mu t1 \\ \lambda_{t1} & -(\mu_{t1}+\mu_{sw}) & 0 \\ 0 & \mu_{sw} & -\mu_{t1} \end{bmatrix} \tag{D.46}$$

$$\begin{bmatrix} -\lambda_{t1} & \mu_{t1} & \mu t1 \\ \lambda_{t1} & -(\mu_{t1}+\mu_{sw}) & 0 \\ 0 & \mu_{sw} & -\mu_{t1} \end{bmatrix} \begin{bmatrix} P_{S_1} \\ P_{S_2} \\ P_{S_3} \end{bmatrix} = \begin{bmatrix} 0 \\ 0 \\ 1 \end{bmatrix} \tag{D.47}$$

$$P = \begin{bmatrix} 0.9994 \\ 2.61 \cdot 10^{-5} \\ 5.49 \cdot 10^{-4} \end{bmatrix}^{tr} \tag{D.48}$$

For option 4:

$$T = \begin{bmatrix} 1-\lambda_{t1} & \lambda_{t1} & 0 \\ 0 & 1-\mu_{sw} & \mu_{sw} \\ \mu_{t1} & 0 & 1-\mu_{t1} \end{bmatrix} \tag{D.49}$$

$$(T-I)^{tr} = \begin{bmatrix} -\lambda_{t1} & 0 & \mu t1 \\ \lambda_{t1} & -\mu_{sw} & 0 \\ 0 & \mu_{sw} & -\mu_{t1} \end{bmatrix} \tag{D.50}$$

$$\begin{bmatrix} -\lambda_{t1} & 0 & \mu t1 \\ \lambda_{t1} & -\mu_{sw} & 0 \\ 0 & \mu_{sw} & -\mu_{t1} \end{bmatrix} \begin{bmatrix} P_{S_1} \\ P_{S_2} \\ P_{S_3} \end{bmatrix} = \begin{bmatrix} 0 \\ 0 \\ 1 \end{bmatrix} \tag{D.51}$$

$$P = \begin{bmatrix} 0.9994 \\ 2.74 \cdot 10^{-5} \\ 5.75 \cdot 10^{-4} \end{bmatrix}^{tr} \tag{D.52}$$

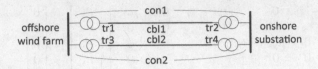

Fig. D.17 Fault tree offshore network 1

4.9 Fault/Event Tree Analysis

Both fault tree analysis and event tree analysis describe in a graphical way how component failures can result in a failure of a system. Fault tree analysis however concentrates on the combination of component failures which leads to a system failure, while event tree analysis concentrates on the series of event that leads to a system failure. Fault trees are drawn vertically and event trees horizontally. Fault trees normally include logical operator symbols and event trees not. Often, two main assumptions are made for the calculations of the probabilities:

- It is assumed that the basic events are independent.
- It is assumed that the probabilities of the basic events are small.

4.10 Fault Tree Analysis

The offshore network and the components are drawn in Fig. D.17. For this question, it is assumed that the offshore network is $N - 1$ redundant. This means that both connections (con1 and con2) must fail to have an offshore network failure.

The fault tree then can be drawn as shown in Fig. D.18.

The probability that the offshore network failed can be calculated by:

$$U_{tr1} = f_{tr1}r_{tr1} = 0.060 \cdot \frac{500}{8760} = 3.43 \cdot 10^{-3} \tag{D.53}$$

$$U_{tr3} = U_{tr1} \tag{D.54}$$

$$U_{tr2} = f_{tr2}r_{tr2} = 0.040 \cdot \frac{48}{8760} = 2.19 \cdot 10^{-4} \tag{D.55}$$

$$U_{tr4} = U_{tr2} \tag{D.56}$$

$$U_{c1} = f_{c1}r_{c1} = 0.010 \cdot \frac{1400}{8760} = 1.60 \cdot 10^{-3} \tag{D.57}$$

$$U_{cbl2} = U_{cbl1} = U_{c1} \tag{D.58}$$

$$P_{\text{failure}} = (U_{tr1} + U_{cbl2} + U_{tr2})(U_{tr3} + U_{cbl2} + U_{tr4}) = (U_{tr1} + U_{c1} + U_{tr2})^2$$
$$= (3.43 \cdot 10^{-3} + 1.60 \cdot 10^{-3} + 2.19 \cdot 10^{-4})^2 = 2.75 \cdot 10^{-5} \tag{D.59}$$

This is equal to $P_{0\%}$ in Example 4.2.

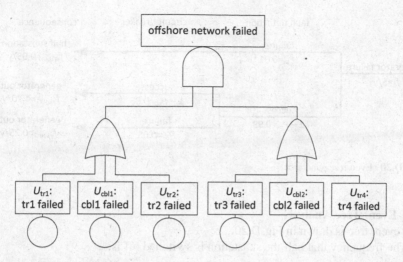

Fig. D.18 Fault tree offshore network 2

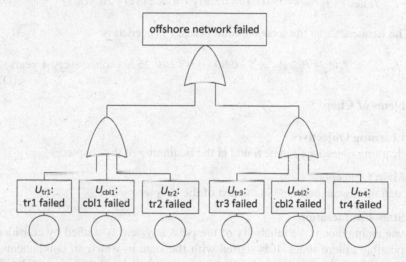

Fig. D.19 Fault tree offshore network 3

If the offshore network is not $N-1$ redundant, the probability of having 50% transmission capacity can be calculated using the fault tree as shown in Fig. D.19:

$$P_{50\%} = (U_{\text{tr1}} + U_{\text{cbl2}} + U_{\text{tr2}}) + (U_{\text{tr3}} + U_{\text{cbl2}} + U_{\text{tr4}}) = 2(U_{\text{tr1}} + U_{\text{c1}} + U_{\text{u2}})$$
$$= 2(3.43 \cdot 10^{-3} + 1.60 \cdot 10^{-3} + 2.19 \cdot 10^{-4}) = 1.05 \cdot 10^{-2} \qquad \text{(D.60)}$$

This is almost equal to $P_{50\%}$ in Example 4.2.

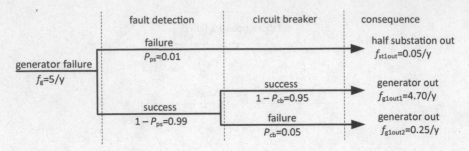

Fig. D.20 Even tree generator

4.11 Event Tree Analysis

The event tree is drawn in Fig. D.20.

The frequency that half the substation is switched off is:

$$f_{\text{st1out}} = f_g P_{\text{ps}} = 5 \cdot 0.01 = 0.05\,/\text{y (=once every 20 years)} \qquad (D.61)$$

The frequency that the second circuit breaker is needed is:

$$f_{\text{g1out2}} = f_g(1 - P_{\text{ps}})P_{\text{cb}} = 5 \cdot 0.99 \cdot 0.05 = 0.25\,/\text{y (=once every 4 years)} \qquad (D.62)$$

Problems of Chap. 5

5.1 Learning Objectives

The learning objectives can be found at the beginning of the chapter.

5.2 Main Concepts

The definitions can be found in the text of the chapter.

5.3 State Enumeration

In state enumeration, the reliability of the power system is studied by considering the possible failure states. It is started with the state in which all components are working. Then, the first level of failure states (first-order states) is considered, in which single components have failed. Then, next levels of failure states are studied, for example second-order states (independent double contingencies or dependent double contingencies). Only, contingencies up to a certain order are considered. This is also illustrated in Fig. D.21.

Also in deterministic contingency analysis, failure states up to a certain level are analyzed. In deterministic contingency analysis however, the probabilities of the specific failure states are not considered. It is, for example, studied whether the system is $N - 1$ redundant, such that single component outages do not lead to any issues in the system. In (probabilistic) state enumeration, the state probabilities are considered. Then, the results of all studied failure states are combined to calculate a reliability indicator like the probability of load curtailment.

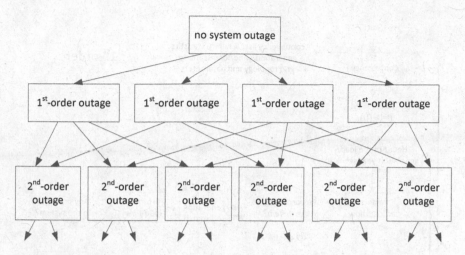

Fig. D.21 State enumeration

As state enumeration is a study of the possible failure state of the system, there are some similarities with Markov models. Also in Markov models, an overview of the possible failure states is made. State enumeration normally does not consider all possible failure states, such that it can be seen as a partial Markov model. The calculation of the state transition frequencies in Markov models can also be applied to calculate the transition frequencies between failure states in a state enumeration.

State enumeration is preferred over Monte Carlo simulation if it is enough to consider the lower-order failure states (i.e. if the probabilities of component failures are small) and if no further information about switching sequences and probability distributions is needed.

5.4 State Enumeration Algorithm

It can be assumed that there are no issues in the system during normal operation. The workflow of this state enumeration is shown in two parts in Fig. D.22a, b. The precise implementation depends on the used software, but the general algorithm is:

% initial values
$P_{error} = 1$ (this will be the total probability of not-considered states)
$P_{redispatch} = 0$ (this will be the probability of redispatch)
$P_{loadcurt} = 0$ (this will be the probability of load curtailment)

% 1st-order: independent single contingencies
for(every independent single contingency)
    P_{state} = calculate state probability
    $P_{error} = P_{error} - P_{state}$
    create the network model
    for(every hour of the year)

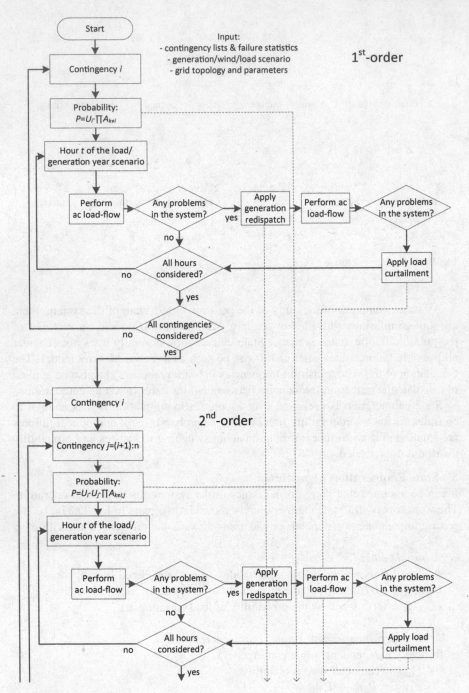

Fig. D.22 State enumeration part 1 (upper part)

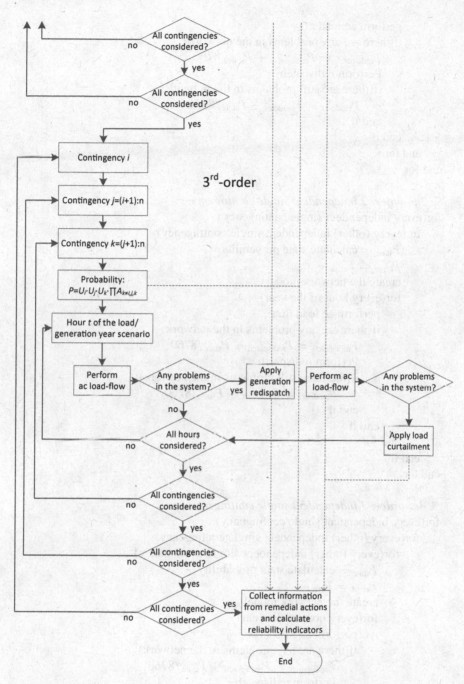

Fig. D.22 (continued)

```
     perform ac load flow
     if(there are any problems in the network)
          P_redispatch = P_redispatch + P_state/8760
          Perform redispatch
          if(there are still problems in the network)
               P_loadcurt = P_loadcurt + P_state/8760
          end if
     end if
  end for
end for
```

% *2nd-order: 2 independent single contingencies*
```
for(every independent single contingency)
  for(every (other) independent single contingency)
       P_state = calculate state probability
       P_error = P_error - P_state
       create the network model
       for(every hour of the year)
            perform ac load flow
            if(there are any problems in the network)
                 P_redispatch = P_redispatch + P_state/8760
                 Perform redispatch
                 if(there are still problems in the network)
                      P_loadcurt = P_loadcurt + P_state/8760
                 end if
            end if
       end for
  end for
end for
```

% *3rd-order: 3 independent single contingencies*
```
for(every independent single contingency)
  for(every (other) independent single contingency)
     for(every (other) independent single contingency)
          P_state = calculate state probability
          P_error = P_error - P_state
          create the network model
          for(every hour of the year)
               perform ac load flow
               if(there are any problems in the network)
                    P_redispatch = P_redispatch + P_state/8760
                    Perform redispatch
                    if(there are still problems in the network)
                         P_loadcurt = P_loadcurt + P_state/8760
                    end if
```

```
                    end if
                end for
            end for
        end for
    end for
```

% results of state enumeration
$P_{\text{redispatch}}$ = probability redispatch in [-]
$8760 P_{\text{redispatch}}$ = probability redispatch in [h/y]
P_{loadcurt} = probability load curtailment in [-]
$8760 P_{\text{loadcurt}}$ = probability load curt in [h/y]
P_{error} = total probability of not-considered states

5.5 Generation Adequacy Analysis

In generation adequacy analysis, it is studied whether there is enough generation capacity to supply the load. It is related to the stress-strength model, as the available generation capacity can be seen as the system strength and the total system load is the stress on the system. The (transmission/distribution) network is not considered in generation adequacy. The available generation and total system load are modeled as probability distributions, as shown in Fig. D.23. Then, the probability that the system stress is larger than the system strength is calculated.

Possible reliability indicators are: *LOLP* (loss of load probability), *LOLE* (loss of load expectation), *LOEE* (loss of energy expectation). For variable renewable energy sources like wind energy, it can be studied how much conventional generation can be replaced by a certain installed capacity of renewables while keeping the same reliability level (*LOLP/LOLE/LOEE*). An example of this is the capacity credit of wind energy.

5.6 Reliability Generation System

This question is comparable to Example 5.7. The COPT can be created in several steps:

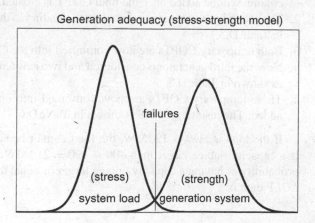

Fig. D.23 Generation adequacy

Generation adequacy (stress-strength model)

failures

(stress)
system load

(strength)
generation system

Table D.2 Initial COPT

C_{out} (MW)	P (-)
0	0.97
100	0.03

Table D.3 Two new (temporary) COPTs

C_{out} (MW)	P (-)	C_{out} (MW)	P (-)
0	$0.97 \cdot 0.96 = 0.9312$	100	$0.97 \cdot 0.04 = 0.0388$
100	$0.03 \cdot 0.96 = 0.0288$	200	$0.03 \cdot 0.04 = 0.0012$

Table D.4 Combined COPT

C_{out} [MW]	P [-]
0	0.9312
100	0.0676
200	0.0012

Table D.5 Two new (temporary) COPTs

C_{out} [MW]	P [-]	C_{out} [MW]	P [-]
0	$0.9312 \cdot 0.95 = 0.8846$	200	$0.9312 \cdot 0.05 = 0.0466$
100	$0.0676 \cdot 0.95 = 0.0642$	300	$0.0676 \cdot 0.05 = 0.0034$
200	$0.0012 \cdot 0.95 = 0.0011$	400	$0.0012 \cdot 0.05 = 6.0 \cdot 10^{-5}$

1. First, a COPT of the first generator is created as shown in Table D.2. The third column will be added once the final COPT is created.
2. The second generator is now considered, leading to the temporary COPTs shown in Table D.3.
3. Both temporary COPTs are now combined into the COPT of Table D.4.
4. Now, the third generator is considered and two new temporary COPTs are created, as shown in Table D.5.
5. These temporary COPTs are now combined into one, and the third column is added. The resulting COPT is shown in Table D.6.

If the load is always 190 MW, the load cannot be supplied completely if there is a capacity outage larger than $400 - 190 = 210$ MW. The COPT shows that the probability of having a capacity outage larger or equal than 300 MW is 0.0034. The $LOLP$ then is:

Table D.6 Combined COPT

C_{out} [MW]	P [-]	$1 - \sum_1^{n-1} P$ [-]
0	0.8846	1
100	0.0642	0.1154
200	0.0477	0.0511
300	0.0034	0.0034
400	$6.0 \cdot 10^{-5}$	$6.0 \cdot 10^{-5}$

$$LOLP = P[C_{out} > 210] = P[C_{out} \geq 300] = 0.0034 \qquad (D.63)$$

The *LOLE* is then:

$$LOLE = 8760 \cdot LOLP = 30.1 \, h/y \qquad (D.64)$$

If the load is 240 MW, the load cannot be supplied completely if there is a capacity outage larger than $400 - 240 = 160$ MW. The *LOLP* then becomes:

$$LOLP = P[C_{out} > 160] = P[C_{out} \geq 200] = 0.0511 \qquad (D.65)$$

The *LOLE* is then:

$$LOLE = 8760 \cdot LOLP = 448 \, h/y \qquad (D.66)$$

If the load is 1/4 of the year 240 MW and 3/4 of the year 190 MW, the *LOLP* is:

$$LOLP = \frac{3}{4} P[C_{out} > 210] + \frac{1}{4} P[C_{out} > 160]$$
$$= \frac{3}{4} \cdot 0.0034 + \frac{1}{4} \cdot 0.0511 = 0.0154 \qquad (D.67)$$

The *LOLE* is then:

$$LOLE = 8760 \cdot LOLP = 135 \, h/y \qquad (D.68)$$

5.7 Monte Carlo Simulation

Monte Carlo simulation is a computer simulation of the behavior of a certain system. Mostly, in the simulation a multitude of years operation in reality is simulated in a shorter time. From the simulation, results like various reliability indicators are calculated. The system behavior in the simulation is thereby used to calculate general characteristics of the system (like reliability indicators or probability distributions).

The accuracy of the results of a Monte Carlo simulation is dependent on the simulation time and not on the complexity of the system. Therefore, Monte Carlo simulation is preferred over state enumeration in highly complicated systems. Also, in Monte Carlo simulation it is possible to include sequential information like switching actions and protection system reaction. Furthermore, with Monte Carlo it is possible to obtain more advanced results like probability distributions.

In sequential Monte Carlo simulation, the time progress is considered. In non-sequential simulation, the time progress is not considered. This means that in non-sequential simulation, sampled values (e.g. load/generation/failures) are independent on the previous samples. Time behavior like protection reaction cannot be included in non-sequential simulation.

5.8 Monte Carlo Algorithm

The workflow of the Monte Carlo simulation is shown in Fig. D.24. The precise implementation depends on the software that is used, but the general algorithm will be:

% initial values
$\text{cont}_{1\text{ind}}$ = list of single contingencies
$P_{1\text{ind}}$ = probabilities of single contingencies
$n_{1\text{ind}}$ = number of single contingencies
$\text{cont}_{2\text{dep}}$ = list of dependent double contingencies
$P_{2\text{dep}}$ = probabilities of double contingencies
$n_{2\text{dep}}$ = number of double contingencies
t_{sim} = 0 (this is the time of the simulation)
$P_{\text{redispatch}}$ = 0 (this will be the probability of redispatch)
$P_{\text{ld.curt}}$ = 0 (this will be the probability of load curtailment)
$E_{\text{redispatch}}$ = 0 (this will be the amount of redispatch)
$E_{\text{ld.curt}}$ = 0 (this will be the amount of load curtailment)

% Monte Carlo simulation
while($t_{\text{sim}} < 1 \cdot 10^6$)
    create a vector $u_{1\text{ind}}$ of $n_{1\text{ind}}$ unit uniform random samples
    $u_{1\text{ind}} < P_{1\text{ind}}$ gives a vector indicating which single contingencies occur
    create a vector $u_{2\text{dep}}$ of $n_{2\text{dep}}$ unit uniform random samples
    $u_{2\text{dep}} < P_{2\text{dep}}$ gives a vector indicating which double contingencies occur
    based on these contingencies, create the network model
    take a sample u from the unit uniform distribution
    the load/generation snapshot is hour ceil($8760 \cdot u$) from the scenario
    perform an ac load flow
    if(there are problems in the network)
        $P_{\text{redispatch}} = P_{\text{redispatch}} + 1$
        perform generation redispatch (E_{rd} = amount of redispatch [MW])
        $E_{\text{redispatch}} = E_{\text{redispatch}} + E_{\text{rd}}$
        if(there are still problems in the network)
            $P_{\text{ld.curt}} = P_{\text{ld.curt}} + 1$
            perform load curtailment (E_{lc} = amount of load curtailment [MW])

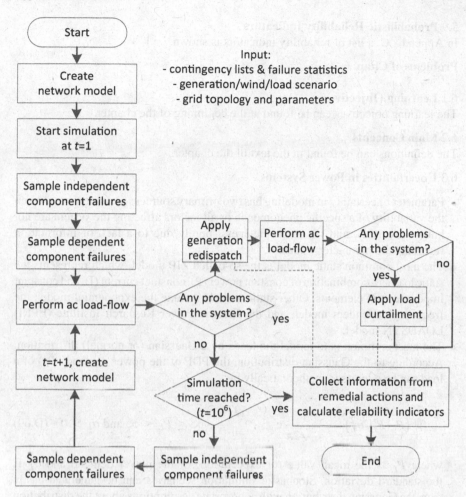

Fig. D.24 Monte Carlo simulation

$$E_{\text{ld.curt}} = E_{\text{ld.curt}} + E_{\text{lc}}$$
$$\text{end if}$$
$$\text{end if}$$
$$t_{\text{sim}} = t_{\text{sim}} + 1$$
$$\text{end while}$$

% simulation results
$$P_{\text{redispatch}} = P_{\text{redispatch}}/1 \cdot 10^6 \ [\text{-}]$$
$$E_{\text{redispatch}} = 8760 \cdot E_{\text{redispatch}}/1 \cdot 10^6 \ [\text{MWh/y}]$$
$$P_{\text{ld.curt}} = P_{\text{ld.curt}}/1 \cdot 10^6 \ [\text{-}]$$
$$E_{\text{ld.curt}} = 8760 \cdot E_{\text{ld.curt}}/1 \cdot 10^6 \ [\text{MWh/y}]$$

5.9 Probabilistic Reliability Indicators
In Appendix C, a list of reliability indicators is shown.

Problems of Chap. 6

6.1 Learning Objectives
The learning objectives can be found at the beginning of the chapter.

6.2 Main Concepts
The definitions can be found in the text of the chapter.

6.3 Uncertainties in Power Systems

- Parameter uncertainty in modeling has two primary sources: (i) randomness, when the variability of a specific phenomenon or all factors affecting the system are not known precisely, and (ii) incompleteness, purely due to a lack of information regarding the parameter values.
- The most common static model is the so-called ZIP model, which represents the static load as a combination of constant power (P), constant current (I), and constant impedance (Z) elements. Other static models include the exponential model, the frequency-dependent model, and the Electric Power Research Institute (EPRI) LOADSYN model.
- The most common is to model the loads by a Gaussian (or normal) distribution. According to the Gaussian distribution, the PDF of the power demand (P_L) of a load can be expressed mathematically by:

$$f_{P_L}\left(P_L | P_L', \sigma_L^2\right) = \frac{1}{\sigma_L\sqrt{2\pi}} e^{-\frac{\left(P_L - P_L'\right)^2}{2\sigma_L^2}} \quad -\infty \le P_L \le \infty \text{ and } \sigma_L > 0 \quad (D.69)$$

where P_L' is the mean value (or expectation) of the power demand and σ_L is the standard deviation. Stochastic simulation scenarios can be generated easily from the Gaussian distribution with appropriate assumptions about the distribution parameters (P_L' and σ_L).

- A Gaussian mixture model is a weighted sum of N_C component Gaussian densities as described by:

$$p\left(\mathbf{x}|\lambda\right) = \sum_{i=1}^{N_C} w_i g\left(\mathbf{x}|\boldsymbol{\mu}_i, \boldsymbol{\Sigma}_i\right) \quad (D.70)$$

where $\mathbf{x}$ is a D-dimensional continuous-valued data vector (i.e. measurement or features), w_i, $i = 1, \ldots, N_C$, are the mixture weights, and $g\left(\mathbf{x}|\boldsymbol{\mu}_i, \boldsymbol{\Sigma}_i\right)$, $i = 1, \ldots, N_C$, are the component Gaussian densities. Each component density is a D-variate Gaussian function of the form:

$$g\left(\mathbf{x}|\boldsymbol{\mu}_i, \boldsymbol{\Sigma}_i\right) = \frac{1}{(2\pi)^{\frac{D}{2}} |\boldsymbol{\Sigma}_i|^{\frac{1}{2}}} e^{\left\{-\frac{1}{2}(x-\mu_i)'\boldsymbol{\Sigma}_i^2(x-\mu_i)\right\}} \quad (D.71)$$

with mean vector μ_i and covariance matrix Σ_i. The mixture weights satisfy the constraint that the sum of all the weights must equal to one:

$$\sum_{i=1}^{N_C} w_i = 1 \qquad (D.72)$$

This condition is because a PDF must be nonnegative and the integral of a PDF over the sample space of the random quantity it represents must be unity.

6.4 Probabilistic Power Flow

- The core of deterministic power flow analysis is solving a set of equations that defines the power balance of the power system (i.e. generation = demand + losses):

$$\text{Power balance equations:} \begin{cases} \sum_{k=1}^{N_{\text{buses}}} P_{\text{gen}} - \left[\sum_{k=1}^{N_{\text{buses}}} P_{\text{load}} + P_{\text{losses}} \right] = 0 \\ \sum_{k=1}^{N_{\text{buses}}} Q_{\text{gen}} - \left[\sum_{k=1}^{N_{\text{buses}}} Q_{\text{load}} + Q_{\text{losses}} \right] = 0 \end{cases}$$
$$(D.73)$$

- Probabilistic power flow analysis can be subdivided into two categories: (i) the numerical approach and (ii) the analytical approach. A combination of both approaches, i.e. a hybrid approach, is possible as well. The analytical approach analyzes the probabilistic power flow considering the inputs as purely mathematical expressions representing the uncertainty. Typical approaches use PDFs as a representation of the input variables, and the output results are obtained in the form of PDFs as well. The numerical approach is to adopt a Monte Carlo (MC) method for the analysis.

6.5 Non-sequential MCS applied to PPF Problem

- A flowchart of the MC simulation mechanism applied to probabilistic power flow analysis is shown in Fig. 6.7 of Chap. 6.
- The probabilistic power flow analysis starts with the set of nonlinear algebraic equations in (6.6) and includes the power system randomness in the power flow formulation by adding this as a component of the input vector (**u**). As a consequence, the input vector is redefined based on two components: one vector (**λ**) containing the deterministic variables (λ_i) and a vector (**X**) containing the random variables (X_i) to model power system uncertainties.

$$\mathbf{u} = [\boldsymbol{\lambda} \ \mathbf{X}]^T \qquad (D.74)$$

- The procedure followed consists of five steps:

  1. *Calculate the correlation matrix (**R**), if not provided*: The matrix **R** of correlation coefficients (r_{ij}) is calculated from an uncertainty matrix $\underline{\mathbf{X}}$ whose rows are observations and whose columns are uncertainties (n). It assumed a unique

correlation between the Gaussian components in uncertainties represented by GMM. If the **R** is known, then go to step 2.

2. *Spectral decomposition.* Spectral decomposition is applied to the correlation matrix (**R**) in order to get two matrices: **D**, diagonal matrix containing the $(n \times n)$ eigenvalues, and **V**, whose (n) columns are the corresponding right eigenvectors.

3. *Generate the multivariate non-correlated normal random matrix.* $\mathbf{X}_{\text{ncorr}}$ is a matrix in which each column is a set of standard normal-distributed generated numbers, $N(0, 1)$, which represent each uncertainty (X_i).

4. *Generate multivariate correlated normal random matrix (*$\mathbf{X}_{\text{corr}}$*).* The mth column of the correlated normal matrix is calculated as:

$$X_{\text{corr},m} = V\sqrt{D}X_{\text{corr},m} \quad m = 1, \ldots, n \tag{D.75}$$

5. *Transformation of the correlated normal random matrix into the desired distribution.* The correlated normal random matrix is transformed into a matrix (**X**) with a distributed random number following desired distributions for each uncertainty:

   a. *Normally distributed uncertainty*: Standardized normal distribution of each uncertainty is reverted to the appropriate mean and standard deviation $N(\mu, \sigma)$.

   b. *Non-normally distributed uncertainty*: The transformation starts calculating the *cumulative distribution function* (CDF) of each uncertainty, and then the inverse probability integral transform or Smirnov transform is used to obtain the desired distribution.

In the case of uncertainties modeled using a GMM, the previous procedure can be applied. However, standardization might be reverted using a procedure related to 5a but considering $\underline{n}$-components and its weights.

Problems of Chap. 7

7.1 Learning Objectives
The learning objectives can be found at the beginning of the chapter.

7.2 Main Concepts
The definitions can be found in the text of the chapter.

7.3 Underground Cable Failure Statistics

- EHV underground cables are a relatively new technology. Therefore, the number of operational EHV cables is too limited to define accurate failure statistics. Failure statistics of underground cables are mainly based on cables of lower voltage levels.
- Cigré TB379 (although this is mainly based on cables of lower voltage levels); the survey among European TSOs as mentioned in the references.

- The failure frequency is based on a survey among European TSOs; the repair time was estimated based on expert's opinion.
- Although the actual repair of a broken cable takes about two weeks, other activities lengthen the total repair time considerably. Especially, logistics (e.g. arranging spare/new parts, availability of technical personnel and permissions) can be relatively long. The repair time can be shortened by optimizing the repair process.

7.4 Underground Cable Circuit Reliability

- Under the assumption that $f \approx \lambda$, it can be assumed that $f_{\text{connection}} \approx \sum f_{\text{components}}$, such that:

$$f_{c1} \approx n_{\text{ic}} l_{\text{total}} f_{\text{cable}} + 3 n_{\text{ic}} \left(N_{\text{cpart}} - 1 \right) f_{\text{joint}} + 3 \cdot 2 n_{\text{ic}} f_{\text{term}} \ [/y] \qquad (D.76)$$

$$U_{c1} \approx n_{\text{ic}} l_{\text{total}} f_{\text{cable}} \frac{r_{\text{cable}}}{8760} + 3 n_{\text{ic}} \left(N_{\text{cpart}} - 1 \right) f_{\text{joint}} \frac{r_{\text{joint}}}{8760} + 3 \cdot 2 n_{\text{ic}} f_{\text{term}} \frac{r_{\text{term}}}{8760} \ [-] \tag{D.77}$$

If all cable parts have the same length (l_{cpart} [km]), the failure frequency and unavailability of a single UGC circuit can be calculated by:

$$f_{c1} \approx n_{\text{ic}} l_{\text{total}} f_{\text{cable}} + 3 n_{\text{ic}} \left[\frac{l_{\text{total}}}{l_{\text{cpart}}} - 1 \right] f_{\text{joint}} + 6 n_{\text{ic}} f_{\text{term}} \ [/y] \qquad (D.78)$$

$$U_{c1} \approx n_{\text{ic}} l_{\text{total}} f_{\text{cable}} \frac{r_{\text{cable}}}{8760} + 3 n_{\text{ic}} \left[\frac{l_{\text{total}}}{l_{\text{sec}}} - 1 \right] f_{\text{joint}} \frac{r_{\text{joint}}}{8760} + 6 n_{\text{ic}} f_{\text{term}} \frac{r_{\text{term}}}{8760} \ [-] \tag{D.79}$$

If all cable circuit components have the same repair time:

$$U_{c1} = f_{c1} \frac{r_{\text{cblcct}}}{8760} \ [-] \tag{D.80}$$

- The failure frequency and unavailability of independent double-circuit failures in UGC connections are:

$$f_{\text{c2ind}} = 2 U_{c1} f_{c1} \ [/y] \tag{D.81}$$

$$U_{\text{c2ind}} = U_{c1}^2 \ [-] \tag{D.82}$$

- To calculate the failure frequency and unavailability of dependent double-circuit failures, a dependent failure factor (c_{cc}) can be defined. The dependent failure factor is the ratio between the failure frequency of dependent double-circuit failures and the single-circuit failure frequency. The failure frequency and unavailability of dependent double-circuit failures then become:

$$f_{c2dep} = c_{cc} f_{c1} \ [/y] \tag{D.83}$$

$$U_{c2dep} = f_{c2dep} \frac{r_{cblcct}}{8760} = c_{cc} f_{c1} \frac{r_{cblcct}}{8760} \ [-] \tag{D.84}$$

7.5 Underground Cable Reliability Improvement

Possible solutions to improve the reliability of an underground cable circuit are:

- Reducing the failure frequency and repair time: By using advanced testing and installation techniques, the failure frequency could be reduced, while the repair time can be shortened significantly by optimizing the repair process.
- Partially cabling: It can be decided to only apply underground cables at those locations where they are the most desirable, thereby limiting the total cable length within a circuit.
- Alternative cable circuit configurations: By installing additional circuit breakers or disconnectors, the failed cable (in the Randstad380 configuration) can be isolated within a limited time such that the cable circuit can continue operation at half its capacity (when acceptable in system operation); by installing a spare cable, the broken cable could be replaced within limited time (although this solution has issues regarding its practical implementation).

7.6 Underground Cables in Large Transmission Networks

The main conclusions can be summarized as follows:

- The impact of an EHV underground cable circuit on the reliability of large transmission networks is strongly dependent on its loading, such that underground cables should be avoided in critical, heavily loaded connections.
- To avoid islanding, critical substations connecting large-scale generation/load centers should not be connected by underground cables alone.
- As the impedance of a connection is of influence on the load flow and reliability of a transmission network, series impedance compensation of underground cable connections must be considered as well.
- The reliability of a cable connection can be improved by installing additional circuit breakers or disconnectors, reducing the failure frequency and repair time of the components, using only one cable per circuit phase (if acceptable), partially cabling the connection or installing less cable sections in the connection.

Problems of Chap. 8

8.1 Learning Objectives

The learning objectives can be found at the beginning of the chapter.

8.2 Main Concepts

The definitions can be found in the text of the chapter.

8.3 Operational Technologies and ICTs for Power Grids

- The OT system gathers data from substation bays and station control systems, e.g. voltage and current magnitudes, active and reactive powers, circuit breaker status, and transformer tap positions, and sends them to the control center. The real-time data is used for power system operation. Furthermore, data is processed by the energy management system for secure system operation, maintenance, planning, and future grid development. System operators control the physical facilities in real time by sending control commands to remote terminal units or station control systems at the substations. Utilities such as transmission and distribution system operators use ICT networks for the non-operational side of the business, e.g. integrated single electricity market, asset and resource management, geographic information system, finance, human resources, and payroll.

- The utility ICT and OT networks have evolved from isolated, monolithic structures to open, networked systems. The physical power and cyber layers have become interdependent. Disruptions in the OT system because of cyber intrusions, increased latencies, and communication failures impact directly the power grid monitoring and control capabilities. Major power system disturbances, like natural disasters, may lead to power system instability, which requires proper control. OT disruptions can impede power system control and implicitly the damping of system oscillations that may lead to cascading failures. Furthermore, cascading events can also be initiated by cyber attacks and multiple cyber-power component failures leading to a blackout. The energy storage devices, such as batteries, and backup power supplies, such as diesel generators, have limited capabilities. In case of an extended blackout, the backup power supplies will eventually run out. Without power, hospitals and other critical services are severely affected. A disruption of the cyber-physical energy system may lead to financial loss, damages, chaos, or even a loss of lives.

- The security controls commonly used to protect OT systems against cyber intrusions are firewalls, intrusion detection systems, virtual private networks, encryption, and authentication.

- Usually, the security controls for OT systems have the same vulnerabilities as security controls used to protect the ICT systems in banking, financial, or commercial institutions. The main issue is that security controls of such ICT systems may be patched, and their vulnerabilities may be mitigated more often than the security controls used for OT systems of industrial control processes. The reasoning is that it is more difficult to update the OT systems of critical infrastructures as this process can affect the normal operation of physical facilities. A disruption of service such as electricity supply to customers may result in regulatory penalties and financial loss. Furthermore, extensive commissioning is needed after each update process that prolongs the voluntary outage for maintenance

8.4 Cyber-Physical Power System Modeling

- The power grid and communication network models are usually decoupled and studied separately. Static and dynamic grid models are used in power system analysis, while the availability and operation of communication networks between various power system controllers, substations, and control centers are assumed to be ideal. With the rapid development of ICT and OT and extensive power grid digitalization, this simplifying assumption is not valid anymore. Integration of power grid dynamic models with large, realistic ICT-OT communication models is essential for the investigation and enhancement of cyber security and resilience to cyber attacks. A comprehensive and integrated cyber-physical energy system model allows the modeling and analysis of complex system-of-system dependencies. Co-simulation of continuous time used for power system analysis and discrete-event systems for communication networks enables in-depth investigations of how cyber attacks at cyber system layer impact business continuity of utilities and integrated energy systems at the physical layer that may result in wide area outages or a blackout. The model of the integrated cyber-physical energy system provides real-time, simulated data from both layers, and it is used for cyber security research and validation of novel methods and tools such as intrusion detection and prevention systems and incident response strategies.

- The CPS layered architecture for a typical power grid includes four layers, i.e. power system, substation OTs, distribution control centers, and transmission system operator. The power system dynamic model is used to compute time-domain simulations, e.g. electromechanical transients, for analyzing power system stability, e.g. voltage, frequency, and rotor angle. The time-domain simulations are synchronized with the operating system clock and computed in real time. The continuous-time electrical parameters such as voltage and current are sampled. The power system layer exchanges simulated data in real time with the cyber system layer during the simulation of power system operating condition. Power grid measurements are computed and reported to the cyber system layer, e.g. voltage and current magnitudes, active and reactive powers, frequency, and status of circuit breakers. If phasor measurement units are considered, the measurement sets include the voltage and current phasors. Control setpoints are received from the cyber system layer, e.g. open circuit breaker, increase power generation, or modify voltage setpoint, and implemented by controllers at the power system layer during simulation run time. The dynamic behavior of the power grid is computed and reported in a close loop. The substation level at cyber system layer includes the OT models of transmission and distribution substations. The communication processes at the bay and station levels are modeled and simulated. The bay control units directly exchange measurements and control setpoints with power devices, e.g. generators, transformers, and power lines. They generate data packets that encapsulate power system measurements. Bay control units are networked with the station control system and communication servers. Data packets are routed through a wide area network and sent to the distribution control centers and transmission operator levels at the cyber system layer. Here, the utility OT and ICT systems are modeled and simulated. The data packets report the power system measurements, which are used for power system monitoring and control. Furthermore, they are used for various system functions and applications such as energy management system, operational planning, grid development, and energy market.

- Queuing theory can be used to model the communication processes and OT infrastructure at the cyber system layer of CPS. It captures the buffering, processing, and data exchange capabilities of each OT device. The OT devices are connected to create local area networks at the substation, control center, and transmission system operator levels and the wide area OT infrastructure. In the OT model based on queuing theory, data packets are received at a λ [packets/second] rate from the power system layer or other OT devices. The buffering capability of the OT device is modeled with a first-in, first-out method for organizing and manipulating the data buffer with a finite queuing capacity K. Each data packet waits in the queuing buffer for a certain amount of time until one of the servers is available to process the packet. The OT processor is modeled by m servers in parallel with a computing power of ρ [packets/second]. Each data packer is delayed in the OT processor another amount of time based on the computational performance of the OT device. In case the m servers are busy processing data packets and the buffer is full, incoming packets are dropped. The OT processing capability is quantified in $m\rho$ [packets/second].

8.5 Cyber Attacks on Power Grids

- A cyber intrusion into the OT system would have devastating consequences that can result in unauthorized control of the power grid, disconnection of power plants and substations from the system, change of control settings, and altering control setpoints sent to transformers and generators. Cyber attacks can lead to excessive torques that damage generators in power plants, overvoltages that damage insulation of power devices, and loss of load. Cyber attacks can also affect the voltage and rotor angle stability of the power system, leading to instability. Simultaneously, multiple generator and line contingencies initiate cascading failures in the power grid and cause a power system collapse. Blackouts in large interconnected power systems are catastrophic failures with a severe socioeconomic impact.
- On December 23, 2015, cyber attacks were conducted against the power grid in Ukraine. Hackers intruded into ICT and OT systems of three electricity distribution companies. Seven 110 kV and twenty-three 35 kV power substations were disconnected from the power grid for hours. The cyber attacks in Ukraine are the first publicly acknowledged cyber incidents to result in power outages that affected 225,000 customers. The hackers shut down power by using phishing emails, BlackEnergy 3, VPN and credential theft, network and host discovery, malicious firmware, and OT hijack. Furthermore, they modified schedules for uninterruptible power supplies, opened circuit breakers, and used KillDisk for wiping of workstations, servers, and remote terminal units.

Index

© Springer Nature Switzerland AG 2020
B. W. Tuinema et al., *Probabilistic Reliability Analysis of Power Systems*,
https://doi.org/10.1007/978-3-030-43498-4